PRAISE FOR
How to Move Like a Gardener

"If you would have a lovely garden you should lead a lovely life," is a Shaker quote long hidden in some deep cerebral crevice that leaped from my right brain as I turned each beautiful page of Deb Soule's *How to Move Like a Gardener*. I nibbled at a few words which engaged my full attention. I couldn't stop. I had to keep reading; then I returned back to each page to look again, to nibble words again. This garden of words, photos, thoughts, and feelings feeds my eyes, my mind, my spirit. I love this book. It expresses the wisdom of experience; tangible weaves with intangible—earth smells, memories, joy—beauty as it is. Here plants and people are not separate kingdoms, classes, and species. Within these pages we know that we are one family from one Mother. We are nature. Here we find the practical instructions on how to lead a lovely life in a lovely garden.

—STEVEN FOSTER, *author, photographer,
and consultant specializing in medicinal and aromatic plants*

Deb Soule is overflowing with the healing wisdom of the plant world distilled through many years of study, experience and observation. Her reverence and respect for nature and deep intuitive capacities are evident in every page of this book. What a gift!

—ROBERT KARP, *Director of
the Biodynamic Farming and Gardening Association*

Herbalist Deb Soule offers the reader a beautifully written, heart-centered gardening book that reads as much like a prayer as it does a practical guide for all gardeners, new and experienced; sure to awaken and inspire one to explore creative ways of tending the rich life held in a well-loved garden.

Weaving stories with sound advise, the author shares wisdom gleaned through years of cultivating not only soil and plants, but also her research and work with the pollinators, bio-dynamic practices and favorite tools used in growing simple herbal remedies to nourish the gardener as they heal the Earth. Honoring traditions, ancestors and the sacredness of carrying on the work of the wise herbalists that came before us, Deb reminds us of the blessing of being called to this work!

—KATE GILDAY, *Herbalist, Woodland Essence*

Way beyond food, flowers, or medicine, gardening as the mindful rhythm in harmony with all living souls.

—C.R. LAWN, *FEDCO Seeds*

How to Move Like a Gardener

Planting and Preparing Medicines from Plants

DEB SOULE

In Memory
of Juliette de Bairacli Levy, Adele Dawson, and Cascade Anderson Geller.

In Gratitude
to the plants and their generous healing gifts.

With a Bow
to the next generation of herbalists and gardeners. May you walk the medicine
path with your heart, hands, and mind open, awake and ready to serve. May
your work be of great benefit to all whose lives you touch.

…be fully absorbed by a garden, not as the individual who created it, but as though it created itself in harmony with cosmic laws, drawing one's inner world and the outer world into balance.

—Diane Ackerman

Contents

*The only way to bring peace to the earth
is to learn to make our own life peaceful.*

—BUDDHA

I

A Gardener's Notebook

A Tibetan prayer wheel at the
entrance to Deb's garden.

PREFACE

The world of nature speaks directly to the human heart.
—Tarthang Tulku

Ever since I can remember I have loved the color green. Long before I knew that a magical woodland garden existed in my home town of South Paris, Maine, I sensed something special was happening behind a particular fence I passed on my way to school. Each morning as I gazed out the school bus window I would strain to see beyond that fence, eager for a glimpse of a flowering plant or a fairy or gnome. Years later, as a young woman, I returned to walk in this garden, which a man named Bernard McLaughlin[1] had begun planting in 1936. On several occasions I had the honor of observing this old and wise gardener at work, quietly bent over a patch of bloodroot or primroses or hauling a wheelbarrow full of weeds.

I never spoke with Bernard directly. I came to his garden to breathe in the sweetness of the 50 varieties of lilacs he had planted and sip the nectar of their delicate flowers. It was here in Bernard's garden that I first began touching plants and paying attention to color, texture, and fragrance. As a young gardener, touch became one of my most important gardening tools, one that with years of daily practice shaped my development as a gardener.

I didn't come to gardening by way of reading books. Like many gardeners, I fell in love with the tasks and patience

---

1   In 1936 Bernard McLaughlin began planting what would become one of Maine's most beloved gardens. A century-old Maine farmstead with massive stone walls and a huge barn provided a unique background for his sophisticated collection of trees, woody shrubs, and perennials.Bernard welcomed visitors to the garden whenever the gate was open, creating a true mecca for garden enthusiasts. With no formal horticultural training, tending the garden single-handedly for most of his life, he eventually became known as the "Dean of Maine Gardeners." Bernard McLaughlin died at age 98 in 1995. Visit the McLaughlin Garden & Homestead in person or online. www.mclaughlingarden.org

required — planting seeds, waiting, watching, watering, listening, praying. I learned, and keep learning, by working and being in the garden, day after day, season after season, and year after year. Thirty-eight years later, as I approach the age of fifty-three, my passion for gardening is deeply rooted in my love of nurturing seeds and soil, watching hummingbirds arrive in spring, feeling a gentle summer rain on my face, and waking to the sweet sound of singing birds.

The early morning songs of birds remind me of nuns and monks chanting. Always in my mind is the memory of when I lived in a small Nepalese village in 1980 and was awakened at dawn by the sound of Tibetan gongs calling the nuns and monks to meditation.

Nepal is where I first encountered herbalists praying and chanting while making medicines, their fingers carefully mixing, rolling, and shaping green- and sepia-colored balls of herbs while they repeated healing prayers and mantras. It was during my second visit to Nepal that an 86-year-old Ayurvedic doctor gave me a prayer to say before collecting herbs. This prayer, which I repeat seven times before gathering herbs, binds me to the sacred and ancient task of collecting and preparing plants for medicine.

As a young woman my interest in gardening coincided with two deeply felt longings. First was a longing to understand what being a woman meant to me. The second was a longing to relearn the ancient tradition of using plants for healing. Noticing the womanly shapes of flowers, burls on trees, and soft mounds of moss awakened my senses. The flavors of fresh rosemary, lavender, and linden flowers intrigued my palate. A conversation with Steven Foster, the herb gardener at a Shaker village in Maine, encouraged me to plant herbs in my vegetable garden. And the gift of Juliette de Bairacli Levy's book, *Common Herbs for Natural Health*, ushered me into the vast and magnificent world of medicinal plants.

My 1978 copy of Juliette's book sits torn and tattered, well-loved, in a library overflowing with books on biodynamics, pollinators, meditation, and herbs. Juliette's devotion to medicinal plants and my own experiences in Nepal inspired me to create Avena Botanicals' healing gardens and herbal apothecary in 1985.

After all these years, it is still the call of a raven, the taste of a bitter root, the glimpse of a hummingbird's shimmering body and the smell of a sweet rose that guide me to pay attention, breathe fully, look deeply, and give thanks for the beauty and mysteries life presents.

ABOVE: Male ruby-throated hummingbird.

BELOW: Collecting Hawthorn flowers (Crataegus phaenopyrum).

OPPOSITE: Basket of freshly harvested calendula flowers (Calendula officinalis).

SOIL

Ultimately, I think, gardening speaks to a deep-seated (deep-seeded?) desire to experience the real, the essential, the astonishingly possible. To garden is gradually to give up control, to fall literally to one's knees and come into closer and closer contact with the tremendous and often bewildering beauty of the living world. Nothing, you find, is at all what you thought it was. Dirt is not dirt, but a teeming mass of microorganisms that turns death back into life.

—JOYCE MCGREEVY

Soil is alive. It breathes. It pulses. It's the precious skin covering the body of Earth, offering protection and allowing plants and trees to take root and grow. Soil calls our attention downward, inward, and homeward. Good-quality garden soil and organic compost smell sweet and earthy. If you have never held a handful of well-tended soil or compost and breathed in its aroma, then please, wait no longer.

In your hand will be many worms and millions of invisible microbes. Bow to them. They're so worthy of our respect.

One of my most beloved gardening teachers, Adele Dawson, said that the compost pile is the most important part of the garden. At age 23 I wondered what she meant. At age 53 I know exactly what she meant. Organic and biodynamic gardeners understand that the yearly addition of compost to one's garden beds contributes

enormously to the quality of the soil: its structure, friability, aeration and drainage, the increase of beneficial microorganisms and earthworms, higher organic matter, thicker topsoil, and better soil respiration. Well-cared-for garden soil allows the roots

of plants to penetrate more widely, giving the root hairs lots of room to develop and in turn producing healthier, stronger, more vital plants. Good-quality soil hangs on the root hairs of plants. This is observable when any medicinal root or root vegetable is unearthed. Over and over again I repeat the same instruction to students: *Be sure to shake as much soil from the roots as possible. Give the soil back to the garden before taking the roots to be washed.*

Rudolf Steiner, the Austrian philosopher and spiritual scientist who initiated the principles of biodynamic agriculture, said that agriculture must be based on enlivening the soil.

In Lecture Five of *The Agriculture Course* Steiner stated " ...vitality must be retained within the realm of the living . . . it should never leave the realm of growth. That's why it is so important for us to recognize that the soil surrounding the growing plants' roots is a living entity with a vegetative life of its own, a kind of extension of plant growth into the Earth." (Creeger and Gardner translation, p. 89)

Skillful gardeners are devoted to the health of their garden soil. Be willing to get down on your hands and knees and smell and hold the soil with the same awe you would bring to holding a hummingbird. Learn about soil by smelling it, reading about it, writing about it, and discussing it with experienced organic and biodynamic gardeners and farmers. Pay close attention to the color, texture, strength, and vitality of your plants. Carefully cultivate and build the soil's fertility. Learn how to water, weed, mulch, prune, and pray. Address plant diseases and harmful insects with a relaxed and alert mind. Encourage the presence of beneficial garden insects and pollinators. Become attentive to using all your senses when in the garden. Always keep the health of the soil in your awareness. And remember that as you cultivate the garden, the garden is cultivating you, both inwardly and outwardly. Gardeners are shaped, pruned, and informed by the garden, moment by moment, day by day, month by month, and year by year.

ABOVE: Corinna transplanting 300 Mexican sunflower (*Tithonia rotundifolia*) seedlings. Their flowers provide nectar for hummingbirds, honeybees, bumblebees, and monarch butterflies.

Holding a male ruby-throated hummingbird.

OPPOSITE PAGE: Avena's gardeners, Lindsey, Tanya, Jillian, and Ray, Preparing garden beds.

Cutting an ashwaganda root.

COMPOST

Fertilizing must consist of enlivening the soil, so that the plant is not in dead soil, and does not have difficulty making it through to the fruiting stage out of its own vitality. The plant does this more easily if from the very beginning it is embedded in something living.

—RUDOLF STEINER

The composting process has always fascinated and mystified me, symbolizing release and renewal, decay and rebirth, and hope and forgiveness. A well-built compost pile recycles and transforms what once was alive into nourishment for the next generation. As gardeners, we participate in all aspects of the birth, death, and renewal cycle, whether creating a new garden bed, thinning greenhouse seedlings, pruning trees, or pulling weeds. Every day we end the lives of so many plants, insects, and microscopic life forms whose names we don't even know. I am forever grateful to the compost pile as a place to lay the day's weeds and food scraps, relieved that the lives just ended will re-emerge as something new.

The garden soil's vitality and health can always be improved upon by yearly applications of compost and biodynamic preparations and by adherence to the additional gardening practices outlined in the sidebar (pp. 12-13). Well-made compost improves the quality and structure of soil and makes available essential nutrients to the growing plant. It maintains and sustains microbiological life and earthworm activity and is an excellent source of humus. Wendy Johnson, the author of one of my favorite gardening books, *Gardening at the Dragon's Gate,* writes:

> The word "humus" has many deep, old roots and connections. In Latin humus means "earth," " ground," and "soil." But it also shares the root for humankind, humane, and exhume, words buried in earth roots. Humble, humiliate, and humorous derive as well from this same fertile root word, as do humid and hubris. I love this richness: it matches humus itself. Yet it is humiliating to know so little about this humid, moist, dark brown culture of humus that supports our life, this culture so humbling in its elasticity and mysterious nature, and so deeply deserving of humane attention. (Johnson, 2008, p. 126)

Like gardening, making compost is both an art and a science, and it requires practice. *Just do it, do it, do it* my gardening mentor Adele Dawson would say. Adele was an awesome gardener, writer, activist, community herbalist, and artist. My favorite painting of Adele's depicts the layers of garden soil over hundreds of years. She painted decomposed plants, fungi, and insect bodies, inviting the onlooker to ponder the incredible communal and diverse biology of compost and soil.

Building an aerobic compost pile

Aerobic compost piles generate heat and steam within three days of being built. Air, water, carbon, and nitrogen are the necessary ingredients for a successfully built aerobic compost pile. Gather your materials—fresh manure, straw, and green plants like comfrey leaves—and choose a permanent site that is well-drained and protected from the wind. Partial shade can be helpful in hot, dry climates.

Whether you're making a long windrow of compost or a backyard three-bin system, use old dried stalks or straw for the bottom layer. This allows for air to enter the pile from below. Compost heaps should not be wider than five or six feet nor taller than six or seven feet. A windrow can be as long as the material and space allows. Each spring and summer at Avena Botanicals we build a few 20-foot-long windrows with the help of my neighbor's tractor. We place the biodynamic compost preparations (described in Chapter II) into the newly built pile within the first three to four days of building it. A month later when my neighbor carefully and slowly turns the pile for the first time, we place a second set of the biodynamic compost preparations into the newly turned pile. It

Additional Gardening Practices

Have a vision for your garden and yourself as the gardener. Dream big and start smaller than you think. Keep the soil's health in the forefront of your mind.

Take time to sit quietly and walk in and around the edges of your garden. This regular practice will help you establish a connection with nature's rhythms and strengthen your inner and outer abilities to be more fully present in your daily life.

Pay attention to the movement of the sun and moon through each month and season, noticing their light and shadows, their ebbing and flowing rhythms, their ever-changing qualities. Contemplate the dynamic forces influencing nature from the larger cosmos.

Avoid compacting the soil of garden beds. Remind your human friends to walk on the paths and avoid stepping directly onto a bed. Learn to laugh and leap over beds whenever the opportunity arises. For long production beds, like 200 feet, make a pathway every 50 feet.

Keep all soil covered with a crop or mulch throughout the year. Avoid leaving soil bare in winter.

Incorporate cover crops and annual crop rotations into your yearly garden plan. Keep regular notes in a garden notebook.

Maintain healthy soil structure by using a rototiller or tractor as little as possible or not at all.

isn't necessary to turn one's compost pile, though turning will speed up the composting process. I highly recommend reading the book *A Biodynamic Manual* in its entirety to better understand building compost and using the biodynamic preparations.

Compost bins can easily be built from recycled wooden pallets that are supported by rebar at each corner. Pound six-foot lengths of rebar solidly into the ground. Then connect the pallets to the rebar with wire. Place the bins in close proximity to your kitchen so you'll use them regularly. We keep a bale of straw next to the bins for placing between the layers of herbal and vegetable scraps, green weeds, leaves, and unbleached paper towels.

Layering the right ratio of carbonaceous (dry and brown) and nitrogenous (green or wet) material is important. Twenty to twenty-five parts of carbon to one part nitrogen is the commonly used ratio. In our 20-foot-long windrow we use freshly cut comfrey leaves and a variety of weeds, straw, dry leaves, and fresh cow manure. The layers of plant material are approximately four to six inches thick and the animal manure eight to twelve inches thick. The first layer placed on top of the 20-foot-long row of dried plant stalks is the fresh cow manure, then dry leaves and straw, and then green leaves. We keep repeating this process

Experiment with single or double-digging or sheet-mulching methods. Avoid digging or using machines during wet or severely dry weather.

Establish permanent beds that are cultivated with hand tools only.

Avoid using chemical fertilizers and pesticides. Read labels and research the so-called organic fertilizers and pesticides before purchasing them. Even substances considered to be "safe" may be harmful to birds and pollinating insects.

Read a book on biodynamic gardening. Apply the biodynamic soil preparation BD 500 at least once in the spring and fall. Apply the silica preparation BD 501 at lease once or twice each growing season. (Refer to Chapter II for information on biodynamics.)

Add the biodynamic compost preparations BD 502–507 into a newly built compost pile. Keep the compost pile covered with straw. If you're unable to make your own compost, find a local farm that does.

Consider spraying your garden with the barrel compost preparation or "Prepared" 500 so that your garden receives the benefit of the six biodynamic compost preparations.

Laugh and dance, sing and pray in your garden.

until the pile is between six or seven feet tall. Once it is built, we cover the entire pile with a thick, six- to eight-inch layer of straw. Each time the tractor turns the pile we cover the top of the pile with new straw.

Hugh Lovel tells a story in his book *A Biodynamic Farm* of being introduced to biodynamics by a wise old man named Peter Escher. Peter told Hugh that the layers in his compost pile were too thick and that "they should never be much over an inch deep for any one layer" (p. 39). This is something I am pondering as I was taught to make the layers at least four inches thick. Hugh suggests using granite dust or basaltic quarry dust (from a local quarry) instead of powdered limestone. "To this end basalt and granite dusts may be better sources of minerals than powdered limestones, as they're more balanced soil building materials" (p. 73). At Avena we're experimenting with sprinkling granite dust on top of each fresh manure layer as we build our compost piles.

As mentioned above, adequate amounts of air and water are necessary ingredients when building a compost pile. Turning the pile a few times aerates it. If there isn't enough moisture in your compost pile, then spray the pile with a hose. Be mindful to not drown all those hard-working microorganisms. Use your intuition. Pay close attention to how your compost pile feels and looks as you're building it. Don't be shy to carefully dig into your pile to access the moisture level soon after it's built, being careful to not expose too much of the pile to the hot sun.

If the pile has too much nitrogen and not enough carbon, a strong smell of ammonia will be emitted. Turn the pile and add some old sawdust or straw to remedy the carbon–

nitrogen ratio. If the pile doesn't have enough nitrogen, then the decomposition process isn't able to happen effectively. Turn the pile and add some buckets of thick manure tea to the compost pile to stimulate the growth of the microorganisms that are needed in the decomposition process.

Use a long compost thermometer, available from seed companies like FEDCO and Johnny's Selected Seeds, to monitor the temperature. For a higher-temperature composting process, required by farms that are certified organic, the temperature of the windrow (also known as a static aerated pile) must ideally rise to 131°F and be maintained at this level for three days. If there is little rise in temperature, a thick manure tea can be added to stimulate the growth of the microorganisms. The high-temperature composting process is known to kill pathogens, fly larvae, and most weed seeds. In his book *A Biodynamic Manual*, Pierre Masson writes,

> Contrary to the practices of conventional or organic composting, the heart of the pile should not get hotter than 130F (55 C). If the temperature rises beyond this, spray the top of the pile. (The disinfection of compost matter takes place through organic processes at a temperature close to 120F [50C], over a period of more than three weeks.) In rare cases where the temperature of the compost does not rise, you have to turn it to get air into the pile and to rid it of excess humidity.

Pierre also writes, "Experience and practice will bring a more detailed knowledge. **Composting is an art:** much like working with (the biodynamic preparations) horn manure (500) and horn silica (501), you have to develop a commitment to the manure and compost."

Well-made compost, rich in humus, is what every garden calls out for. Creating and supporting the biological processes in soil with compost that is teeming with beneficial microbes grows vibrant and healthy plants and lowers the potential for pest problems and plant diseases. May more and more backyard gardeners, restaurants, colleges, universities, and municipal communities transform their vegetable "waste" into valuable compost and their lawns into food and herb gardens!

Carbon-to-nitrogen ratio: 25:1

The following ratios of materials used to build a compost pile ensure the appropriate carbon-to-nitrogen ratio.

Green plants and weeds 20:1	Straw 40–80:1
Kitchen scraps 15:1	Well-rotted litter straw 30:1
Manure 15:1	Unbleached paper 200:1
Grass clippings 15:1	Sawdust 500:1
Dry leaves 60:1	

(Adapted from *Grasp the Nettle, Culture and Horticulture,* and *Gardening at the Dragon's Gate.*)

TOOLS

Our favorite tools are an extension of ourselves, almost reaching for the hand as we grasp them, and becoming a perfect marriage of function and form as they're used.

—MONTAGU DON

Tools are to be treasured, wiped clean, sharpened, oiled, and used. They deserve to be hung and stored in a place as protected and sturdy as your grandmother's hand-made china cabinet. That may sound outlandish, but on most farms, garden tools are used far more often than china.

No one showed me how to properly hold or sharpen a spade or efficiently use a hori-hori for transplanting when I was a young gardener. I learned and keep learning by observing gardeners using tools unfamiliar to me. An excellent book for reading about tools is *The Garden Primer,* written by Barbara Damrosch.

These are the garden tools that live in Avena's barn:

Digging and planting tools:

border fork	mattock
digging fork	transplanting spade
edging tool	EZ-Digger
asidon	hori-hori

Border forks are similar to digging forks but smaller in size. An asidon is used for creating new garden beds and hillside swales. A hori-hori is my favorite tool for transplanting seedlings, for prying out deeply-rooted weeds, and for cleaning off the soil from roots and weed clumps.

Tools for lifting and loosening the soil:

border fork	manure fork
digging fork	potato fork
round-pointed shovel (long-handled)	

Julia Yelton using a mattock

Cultivating and weeding tools:

border fork	potato fork	cobra-head weeder
garden hoe	scuffle hoe	hori-hori
EZ-Digger	small mattock	hand fork

You should also have five-gallon plastic buckets for carrying weeds, as well as a wheelbarrow for hauling weeds and distributing compost around the garden.

Pruning tools:

Felco pruning tools are my favorite. They're made in Switzerland from high-quality steel and forged metal alloys with ergonomic designs. They're easy to maintain and have replaceable parts. FEDCO's catalog offers many varieties for sale. I use the Felco #6 pruners for cutting and pruning thick herb stalks, grape vines, shrubs, and young trees; Felco's #21 lopper (has a longer handle) when cutting branches up to one and a half inches in diameter; and Felco's #600 folding pruning saw when pruning fruit trees and when cutting branches (up to four inches in diameter) from shrubs such as the high-bush cranberry. I scrape the early spring bark from the high bush cranberry shrubs, also known as cramp bark*, with a hand-made knife and tincture it fresh for medicine.

*The botanical name of the cramp bark growing on Avena's farm is *Viburnum trilobum*.

Raking tools:

bamboo rake (small and large)	level-head rake

My favorite level-head rake was made in England from steel. It has a long handle and flat top, and I use it when raking and preparing new garden beds.

Herb-collecting tools:

Felco #310 harvest shear	Joyce Chen scissors	kama

I use the harvest shear and the scissors for harvesting most of Avena's medicinal leaves and flowers, which amounts to about sixteen hundred pounds of fresh herbs per year.

A kama is a wooden-handled tool with a flat steel blade that can be kept sharpened with a composite steel blade or Diafold's folding flat file (available through FEDCO). We use a kama when harvesting two hundred pounds of fresh nettle leaf in the springtime.

Kama   Hori-hori   Felco shears

Tools for biodynamic preparations:

Here is what you'll need for making the BD preps: four-gallon ceramic crock or copper pot for stirring biodynamic preparations; clean food-grade five-gallon plastic buckets for carrying around the garden when applying the BD 500 soil preparation; a wooden-handled wallpaper brush for applying the BD 500 soil preparation, and a backpack sprayer (2.5-gallon) for applying the BD 501 silica preparation.

Watering tools:

For best watering, you'll need good-quality rubber hoses with brass couplings, y-spigot, and heavy-duty watering can. Look for lead-free hoses, as they're becoming more available. A brass hose y-spigot with valves allows you to connect two hoses into your main hose—these come in plastic, but beware, the plastic ones break easily. I prefer a long-spouted watering can that allows me to easily water seedling trays in the greenhouse, hanging plants, and window boxes.

The first digging fork I ever purchased was from an old funky junk shop. Within the first week of using this fork the wooden handle broke. At that point I realized that there's a difference between cheaply made tools and the more expensive, well-made ones. I began reading gardening catalogs and purchased my first border fork from England. I still have this border fork. It's one of my favorite tools along with my hori-hori, kama, Felco #6 pruners and Felco #310 harvest shears. Pay attention to the size and weight of a tool, what kind of metal or wood it is made from, where it is manufactured, the cost, and even its color. Tools can easily get misplaced and lost in the garden. Brightly painted handles and a tool bucket often prevent the loss of a precious tool.

I am a hands and knees gardener. Throughout the growing season my fingernails contain bits of beautiful brown earth and the fine lines of my hands are permanently soil colored. My knees are still flexible thanks to yoga and Solomon's seal root tincture and oil. The soles of my feet are thick and sturdy from my early-morning ritual of walking barefoot through the garden. Always in my back pocket are a pair of Felco shears. In a nearby bucket a hori-hori for weeding and an EZ-Digger for cultivating await the familiar touch of my right hand.

My favorite thing to do at the end of a long gardening day is to climb into the old claw-foot bathtub[2] that sits at the edge of my garden and soak amidst floating flowers. In June, whole peonies and roses float in my bathwater, delighting my senses and softening my skin. As the summer progresses, the bathtub shimmers orange and green with calendula blossoms and lemon balm leaves. Water, fire, fragrance, and earth. I speak your names aloud each evening as I bow to you in gratitude.

---

2 The way to easily fill an outdoor bathtub with hot water is to run a hose to a nearby outdoor spigot that is plumbed for hot and cold water.

BASKETS

We are the Wabanaki, the People of the Dawn. Comprised of the Maliseet, Micmac, Passamaquoddy, and Penobscot, our home is in the east where the rising sun first greets the lands of what are now known as Maine, New Brunswick, and Nova Scotia. Our traditions tell us that we have always been here. In one of our Creation stories the people were literally born from the ash tree-the basket tree-dancing and singing. For at least 12,000 years, our people have been in this land creating beautiful objects from the resources surrounding us.

—Jennifer Neptune, *Penobscot basket maker*

Herbalist Mary Bove, one of my earliest teachers, began teaching me how to collect medicinal herbs when I was a teenager. She instilled in me a deep respect for the practice of stopping and communicating with a plant and offering a gift of tobacco before harvesting its flowers, leaves, roots, bark, or seeds. I don't remember exactly when, but early on I realized that to place medicinal plants into hand-made baskets is an honorable way to acknowledge each plant's Spirit and healing gift.

I spent my early life around a grandmother who wove cloth on a small loom and created jewelry from stones she collected and polished. My grandmother Katherine passed on her appreciation for hand-crafted and functional art pieces to me. Many years after her death I purchased my first basket woven by Passamaquoddy basket maker and herbalist Fred Tomah. This basket became the first of many created by Passamaquoddy and Maliseet basket makers that I acquired for collecting herbs. The handles of these baskets are stained from the plant oils that have penetrated my hands as I have collected and placed the plants into these sturdy and beautiful baskets over the past 25 years.

OPPOSITE PAGE: My first student, Rebecca Willow, carrying a basket full of lady's mantle flowers.

ABOVE: Jennifer Neptune, Penobscot basketmaker, collecting sweetgrass (*Anthoxanthum nitens*).

21

There are many stories woven into a basket, beginning with the life of the ash tree, the birds and the insects who depend on the tree for food and shelter, the billions of soil organisms who support the tree's life, the community of plants and trees living nearby, the person who carefully selects and cuts the tree and strips and pounds the bark, the person who weaves the basket, and the people who teach the next generation of basket makers. Within the hands and heart of every basket weaver are generations of experiences, wisdom, and prayers.

In an essay entitled *Spirit of the Basket Tree*, Penobscot basket maker Jennifer Neptune concludes her piece with the following two paragraphs:

> From pack baskets tucked in the curves of our birch-bark canoes to exquisitely woven hex-weave sifting baskets, to the utility baskets peddled to farmers and fishermen, to the wild and colorful fancy baskets of the Victorian era, to the works of art produced today, baskets have played an important role in the lives of the Wabanaki. The baskets reveal not only the spirit of the basket tree but also the spirit of the people—our resistance to assimilation, our creativity, our hopes, our sovereignty and strength.

> Baskets remain a connection to our past, an unbroken chain of mothers, fathers, grandparents, and ancestors who protected and passed on the tradition for those who would need it in the future. Baskets hold memories of family and tribal stories of those who came before. Baskets are a connection to the spiritual, as many weavers still dream their basket designs and feel the hands of the ancestors guiding their work. Sweetgrass still

carries our prayers to the Creator and reminds us of the promise of summer while weaving on cold winter nights. Baskets hold our communities together by connecting young to old, and weavers to apprentices, ash pounders, block and gauge makers, sweetgrass gatherers and braiders, and other weavers and friends in our own tribes and beyond. More than a means of survival, baskets continue to be an inseparable part of the culture and traditions of the Wabanaki.

The purchase of my first basket alerted me to how little Native American history I was exposed to while growing up in Maine. The lack of understanding and respect for indigenous peoples worldwide permeates government agencies and institutions, rural and urban communities, and school systems. I have included suggestions for books, documentary films, and educational forums in the Resources section for readers to utilize and better comprehend the complex issues Native peoples face today both in North America and internationally. Also included is information on how to purchase Native American baskets.

ABOVE: The start of a traditional sweetgrass basket by Jennifer Neptune.

BELOW: A traditional brown ash gathering basket with self-heal flowers.

OPPOSITE PAGE: Gretchen sorting through baskets filled with different medicinal herbs.

POLLINATORS

Last night, as I was sleeping,
I dreamt—marvelous error!—
that I had a beehive
here inside my heart.
And the golden bees
were making white combs
and sweet honey
from my old failures.

—Antonio Machado, *translated by Robert Bly*

Bees. Beetles. Bats. Flies. Wasps. Thrips. Skippers. Moths. Butterflies. Hummingbirds. Hymenoptera. Coleoptera. Diptera. Lepidoptera. Thysanoptera. Chiroptera. Apodiformes. Odonata. I have a crazy fascination with attempting to pronounce the scientific names of plants and pollinators, including their family names and orders. This obsession has something to do with a longing to know my ancestors and a desire to write and speak on behalf of pollinators and plants.

The Hine emerald dragonfly belongs to the Order Odonata and is on the list of endangered insects in the continental United States. So are 12 beetles (Order Coleoptera) and 20 butterflies (Order Lepidoptera). Seven bee species are in decline in North America. A book I turn to often, *Status of Pollinators in North America*, published in 2007, lists 50 insects that have recently become extinct around the world. In the Preface of this book is written:

> Of the multitude of ways humans could be harming the planet, however, one that has largely been ignored is the "pollinator crisis"—the perceived global decline in the number and viability of animal species that facilitate reproduction of flowering plants, the overwhelming majority of plants in terrestrial communities.

Before Rachel Carson, scientist and author of *Silent Spring*, passed away in 1964, she observed that *many herbs, shrubs, and trees of forests depend upon native insects for their reproduction; without these plants many wild animals and range stock would find little food. Now clean cultivation and the chemical destruction of hedgerows and weeds are eliminating the last sanctuaries of these pollinating insects and breaking the threads that bind life to life.*

I try and pause each time I hear or read about another species whose life is in danger or whose life has gone extinct. I often ask myself: *How do I not close my heart from these devastating truths?* Buddhist teacher Tara Brach says: *The very nature of our awareness is to know what is happening. The very nature of our heart is to care.* I don't have answers to the seemingly insurmountable problems of these times. I keep walking back into the garden for instruction and planting habitats for the hummingbirds and bees.

Pollinators are animals, and this kingdom of animals called *Animalia* includes insects that both visit and effectively move pollen from the anthers of one flower to the stigma of another. This movement of pollen, essential to the life cycle of flowering plants, is called pollination. It's estimated that over 85% of the world's 352,000 flowering plants rely on at least one of the 200,000 species of animal pollinators. Once pollen is transferred, fertilization occurs and the resulting seed holds within it the miracle that will become the next generation.

Many of us, including gardeners, rarely stop and contemplate all the ecological and mystical processes that happen to the seed we hold in our hand, the fruit we are tasting, or the herb we are making into medicine. I am embarrassed to admit that I had been gardening for 15 years before I began to consider the significance of pollinators. Without their presence on our planet, life would cease to exist as we know it.

Biologist and author Gary Paul Nabhan writes in *Cross-Pollinations: The Marriage of Science and Poetry:*

Hands holding sweet cicely (*Myrrhis odorata*) seeds, and a Figwort flower (*Scrophularia nodosa*) with bumblebee.

I have learned in my heart what extinction is all about, especially the extinction of relationships. I have touched with my own hands plants so rare, so threatened, that they're no longer able to attract the pollinators they need to set seed and pass on their genes. Extinction seldom comes in one fell swoop, with a bulldozer's scoop or the shot of a single gun barrel. Instead, it occurs when a web of supporting relationships unravel. It occurs whenever we or any other species are unable to sustain mutually beneficial interactions with those around us, those with whom we have been historically associated. No, few endangered plants and animals suffer their ultimate extirpation by being physically removed from the earth. They die by suffering from the loss of ecological companionship.

Every spring I eagerly await the arrival of the ruby-throated hummingbirds to my garden. They're my favorite pollinator. I cannot and do not imagine my garden without their presence. In winter, I dream of them in the warm tropic. In summer, I observe the medicinal plants and trees they frequently visit. In late summer, in preparation for their 3,000-mile migration, I make sure there is an abundance of flowering phloxes, nasturtiums, asters, and salvias for their nectar and pollen needs. When my students ask: *How can I make a difference in these challenging times,* I respond with the words of my favorite writer, Terry Tempest Williams:

Engage. Participate. Love. Any of these actions of the heart will lead to a personal transformation that bears collective gifts.

mon bergamot (*Monarda citriodora*).

Nasturium (*Trapaeolum majus*).

Hummingbird sage (*Salvia coccinea*).

ANCESTORS

The temple bell stops but I still hear the sound coming out of flowers.
—BASHO, 1690

There is a specific place in my garden where I go to talk with the Ancestors. Here I sit quietly and wait to feel their presence. I come here for connection, for conversation, for comfort, for guidance, for spiritual instruction, for inspiration, for renewal. I come here to pray, to meditate, to remember, to observe, to sing, to offer incense, to be still, to say thank you. In this place the Ancestors remind me that we are never alone, that all of life is exquisitely interconnected, and that the Ancestors are sitting with us every day.

GRATITUDE

The ultimate goal of farming is not the growing of crops but the cultivation and perfection of the human being.

—Masanabu Fukuoka

In 1982, while traveling in Japan, I spent three days with Masanabu Fukuoka, author of *The One-Straw Revolution*. While observing Mr. Fukuoka work, listening to him speak, and reading the unpublished translations of his next two books, *The Natural Way of Farming* and *The Road Back to Nature*, I came to recognize that farming had become a spiritual practice for him as it was becoming for me. Though different than a meditation hall, Avena's garden has the feel of a sacred sanctuary. Inside its borders of hawthorn trees and stone walls my body and mind keep learning how to be in rhythm with nature's rhythms. Here I practice mindfulness while collecting hundreds of tiny mullein flowers at dawn or while standing quietly for long periods of time observing ruby-throated hummingbirds. In the evening I sit silently under the stars, remembering Storm Jameson's words, "There is only one minute in which you are alive, this minute here and now. The only way to live is by accepting each minute as an unrepeatable miracle."

At dawn and before bed I light incense and offer prayers of gratitude for the day's lessons and the plants' medicine. Most mornings I meditate quietly at the edge of the garden with a cup of sacred basil tea. On occasion I forget to drink my tea and instead race further into the garden, eager to see what has changed during the night. The garden continues to teach me how to show up, and be present, whatever the internal and external weather may be. It's a place where gratitude lives.

PLANTS AND PRAYERS

There are many ways to pray, the Grandmothers say, and no right way. A prayer can be made in any moment we find ourselves. Their only advice is that people pray with gratitude, and pray for all of Creation, since we are all a part of Creation.

—CAROL SCHAEFER

My grandmother Katherine never spoke to me about prayer. I just felt her quietly praying while we sat and observed the emerging spring flowers and migrating birds. As a child, I was still connected to the world of fairies and flowers, and my grandmother's silent and sacred pauses easily entered me. Praying with plants became familiar and natural. I am incredibly grateful to my paternal grandmother for inviting me along on her spring excursions. Perhaps she recognized the unspoken hunger in me for connection with the natural world. Here, when I walk in the woods or sit on the edge of a moss-covered stream bank, I feel her presence.

My maternal grandmother, May Preti, was raised in the Christian Science tradition. Prayer was central to both her and my mother whenever illnesses arose in our family. It was rare that antibiotics or other pharmaceutical medications were used when a family member fell ill. I experienced only one ear infection as a child. My mother kept me in bed and read aloud to me, fed me warm soup, used a thermometer to monitor my temperature, and had faith that my body knew how to heal.

Approaching a plant with reverence is akin to praying in that both activities invoke mindfulness. Before harvesting a plant intended for medicine I kneel near the plant and repeat the prayer given to me in Nepal. Sometimes I speak my prayer aloud, sometimes silently. I communicate with the plant why I am asking for its medicine, offer a gift, and wait to feel the plant's response before picking up my digging tool or clippers. If the plant's response is "no, it is not the time for you to harvest me," then I quietly bow to the plant, gather up my tools, and move to a different gardening task.

My experience is that plants communicate with us when we take the time to create a relationship with them. Approach plants and trees as you would a friend. Begin by sitting quietly near a plant or tree you are drawn to and notice what sensations arise for

you. Visit this plant or tree as often as possible. Bring simple gifts—a prayer, poem, song, or tobacco—and with hands held in a prayerful gesture offer gratitude and delight for your connection.

Over time, and with repeated effort, the external veil of the natural world will be drawn aside. "We can find our way into this world by means of meditative exercises that allow the sense impressions of nature to live vividly within us and then dissipate, leaving behind a certain soul development, a feeling of inner harmony with nature" (Marcia Merryman Means, *What Is Biodynamics?*, 2005, p. 45).

And as Rudolf Steiner stated in his lecture on Spiritual Beings I: "All the rest will, in certain respects, come of itself when, as time goes on, we acquire an understanding born of feeling and perception for the fact that behind our sense world, behind the world that we as human beings experience, there lies a world of spirit, a spiritual world."

The Medicine Buddha's chant is one I often recite while transplanting seedlings, collecting herbs, and preparing medicine. This chant invokes the healing Buddha to be present and to remove the pain of illness and spiritual ignorance. Chanting for several minutes or for an hour or more helps me be fully present with the sacred task of preparing plant medicines.

Tayata Om Bekandze Bekandze Maha Bekandze Radza Samudgate Soha

We are here to awaken from the
illusion of our separateness.
—Thich Nhat Hanh

II

Biodynamics: Agriculture in Service of the Earth and Humanity

INTRODUCTION

Biodynamic agriculture is a way of living, working, and relating to nature and the vocations of agriculture based on good common-sense practices, a consciousness of the uniqueness of each landscape, and the inner development of each and every practitioner.

—from the journal *Biodynamics*

In 1986, I purchased my first *Stella Natura Biodynamic Planting Calendar*. This calendar introduced me to the biodynamic system of gardening and farming that relates the ecology of the Earth to the larger cosmos, views the soil and farm as living organisms, places significant attention on soil health, consciously works with seasonal, solar, and lunar rhythms, and recognizes and nurtures the spiritual presences and unseen life forces active in the garden. Having been inspired by the book *The Findhorn Garden* when a teenager, I was grateful to find an ecological and holistic approach to agriculture that included an understanding of elemental beings, planetary influences, and the spiritual life of the farmer and farm.

The principles of biodynamic agriculture are outlined in a series of eight agricultural lectures given by Dr. Rudolf Steiner in Koberwitz, Silesia, June 7–16, 1924. Dr. Steiner agreed to lecture on agriculture after a group of farmers approached him for advice regarding the diminished viability of their seeds and decreased vitality of their livestock—observations that coincided with the end of World War I when left-over munitions were beginning to be used as agricultural chemicals. The farmers who came to Steiner were familiar with Steiner's spiritual philosophy known as Anthroposophy (*anthropo* = "human being," *sophia* ="wisdom, the inherent wisdom of humanity"), a philosophy influenced by his own clairvoyant view into spiritual realms—realms beyond the physical world. The eight agricultural lectures Steiner presented in 1924 are

rooted in Anthroposophy.[3] An English version of the eight agricultural lectures, translated by Catherine E. Creeger and Malcolm Gardner in 1993, is available in a book titled *Agriculture*.

Steiner viewed Earth as a living organism and believed that the environmental damages caused by chemical fertilizers, toxic pesticides, and intensive farming methods could not be easily remedied. His ideas for helping heal the Earth were influenced by his early life in rural Austria, where he grew up around people who followed the rhythms of nature and relied upon herbal medicine for healing. Steiner's formal education included philosophy and the scientific work of Johann Wolfgang von Goethe.

Though Dr. Steiner passed away in 1925, the seeds of biodynamic agriculture had been planted, and a group of farmers who called themselves the Experimental Circle began the task of working with and evaluating the biodynamic principles.

One of the core beliefs and practices of biodynamics is to help heal the Earth.

A biochemist named Dr. Ehrenfried Pfeiffer, who had worked closely with Dr. Steiner during the early stages of biodynamics, came to speak at the first biodynamic conference held in the United States in 1933. In 1938, the Biodynamic Farming and Gardening Association (BDA) was formed, becoming the first organic farming organization in the United States.

Today biodynamics is practiced in over 40 countries. It's spreading in India, and in Switzerland, Germany,

---

3   Anthroposophy is described by some practitioners as *the consciousness of one's humanity*.

and Austria, biodynamics is recognized by government agriculture departments as a valuable ecological system of agriculture. In Warmonderhof, the Netherlands, there is a well-established biodynamic training center, and there are training programs in Switzerland, England, Germany, North America, and New Zealand.

The North American Biodynamic Apprenticeship Program (NABDAP), sponsored by the Biodynamic Farming and Gardening Association, offers a two-year training program for North American students interested in biodynamics. In addition to sponsoring this official training program, the BDA encourages bioregional study groups, apprenticeship opportunities, and biodynamic prep-making events throughout North America. The BDA publishes a quarterly journal in which it defines itself as an "association of individuals and organizations in North America who are committed to the transformation of the whole food system, from farm to table, and who draw inspiration from the spiritual-scientific insights of Rudolf Steiner." Information on the BDA is listed in the Resources section.

ABOVE: Male goldfinch sitting atop an Echinacea purpurea flower.

BELOW: Bridget and Kiira learning to collect elder flowers (*Sambucus canadensis*).

A RENEWAL OF AGRICULTURE

The solution to the food crisis is to reclaim food sovereignty and rebuild local food economies based on ecological farming. This path also frees agriculture from its dependence on fossil fuels while increasing mitigation and adaptation to climate change. A shift from oil to soil addresses the triple crisis of climate, energy, and food.

—Vandana Shiva

A new agricultural paradigm—one that shifts from the industrial mind-set of agribusiness to models that value biodiversity, cultural diversity, and ecologically sound agricultural practices—is needed *now*. We have a tremendous amount of challenging work ahead to place caring for the Earth at the forefront of our activities and to ensure that all people and animals have access to high-quality, healthy food. People-led initiatives such as The International Council of the Thirteen Indigenous Grandmothers, The Slow Food movement, and the Community Supported Agriculture (CSA) movement are inspiring thousands of people in North America and Europe. Writers and lecturers like Vandana Shiva, Winona LaDuke, Wendell Berry, and Frances Moore Lappé are encouraging people to rethink how they live and how much they consume.

In the midst of the cultural and ecological horrors of our time, I find myself again and again turning to the study and practice of biodynamics, herbalism, and meditation. I am a small-scale biodynamic farmer and herbalist, tending three acres of medicinal plants and preparing herbal remedies for my community. The ancient and sacred act of planting seeds, nurturing plants, and helping people and animals heal through the use of gentle herbs keeps me directly connected to the natural rhythms of the Earth and

my intention to be of service. The spiritual foundation of biodynamics adds depth and meaning to my life.

Any biodynamic practitioner will tell a curious organic gardener that biodynamic gardening embraces similar ecologically based organic gardening practices. The use of compost, green manures, cover crops, crop rotations, companion plants, and mulch are some of the practices used by both systems to build healthy soil and grow healthy plants. Creating and maintaining fertile soil without the use of toxic chemicals is fundamental to all biodynamic and organic farms and gardens. Ecologically based gardening and farming methods have been practiced by traditional people around our planet for several thousand years, long before the terms *organic gardening* and *biodynamics* were coined. What is different today is that we live on a planet with over seven billion people to feed. The importance of ecologically based farming practices and access to fresh and local food for everyone is not a top priority for most governments. We face new and numerous environmental crises concerning water, climate, toxic pollution, poverty, health care, and food safety and security issues. Over 90 years ago Steiner said, "There is no realm of human life that is not affected by agriculture."

ABOVE: Hand-harvesting green milky oat seeds at Hope's Edge Farm for Avena's tinctures and teas. Lauren gathering sacred basil leaves and flowers. A dew-covered Mexican sunflower.

OPPOSITE PAGE: Avena's hoop house overflowing with seedlings.

It's in the spiritual realms that biodynamics goes beyond organic agricultural practices. Steiner taught that behind all physical substance is a spiritual impulse. This spiritual impulse connects the material world with the larger cosmos and with the unseen life forces that give energy and physical form to minerals, animals, plants, and humans. One of the core beliefs and practices of biodynamics is to help heal the Earth. This is accomplished through the use of specially made, energetically dynamic preparations that are applied to the soil and compost pile following the natural rhythms of the day, the seasons, and the movement of the moon and planets. Steiner taught that the regular use of the biodynamic preparations (described below) enlivens the soil and allows it to be more open and receptive to the spiritual impulses and beneficent forces radiating from the larger cosmos. He believed that the vitality of the food and herbs we ingest, revitalized by the use of the biodynamic preparations, would help humans become more fully conscious, more spiritually awake.

In the introduction to the book *What Is Biodynamics? A Way to Heal and Revitalize the Earth,* Hugh J. Courtney writes:

> Humanity is not able to grow spiritually because the food that nourishes the human being no longer carries the forces that will lead to the necessary spiritual development. Steiner, in a conversation with Ehrenfried Pfeiffer, identified humanity's lack of spiritual development as "a problem of nutrition." The ultimate solution to the problem of nutrition and the lack of spiritual development lies in finding the means to restore forces to nature that will then restore truly healthy life to the Earth. The means to do this were given by Steiner in the Agriculture Course, "so the earth may be healed."

BIODYNAMIC PRACTICES FOR
SUPPORTING THE EARTH'S HEALING

The word biodynamic combines two aspects of growing: BIO which refers to bringing life into matter and DYNAMIC which suggests the rhythmical movements (day, night, seasons). The combination of the two forces results in the dynamic biological growing activity known as biodynamic agriculture.

—BRIAN KEATS AND STEFAN MAGER

The following biodynamic practices are ones I have incorporated into my gardening and farming activities for restoring the Earth's vitality. These practices have enhanced my inner capacity for understanding and valuing the natural processes that support Earth's life. The creation and use of the biodynamic preparations may seem odd until one understands and honors how integral the mineral, plant, and animal realms are in enhancing the ecological and spiritual health of Earth. Biodynamic farming offers humans a path to develop our inner capacities for helping our planet to be a place for life to flourish.

ABOVE: Seed pods of the poppy.

1. Regularly applying the biodynamic preparations, including BD 500 (horn manure) and BD 501 (horn silica), the six compost preparations (BD 502-507), horsetail (BD 508), barrel compost, the Three Kings prep, and the clay and cow dung prep for trees. I call these preparations Medicines for the Earth.

2. Using the biodynamic planting calendar as a guide for seeding, transplanting, and harvesting activities. Consciously working with the lunar, solar, and

47

seasonal rhythms, the movement of the planets, and the four elements: earth, air, fire, and water.

3. Attending to soil health through the use of biodynamically prepared compost, BD 500, crop rotation, cover cropping, and mulching.

4. Recognizing and communicating with the spiritual presences and unseen life forces active in the garden, farm, forests, and wilderness. Honoring the ancestors and the nature spirits and elemental beings associated with place, plants, trees, birds, pollinators, and the four elements.

5. Closely observing the natural world.

6. Viewing the farm as a living organism and the gardener and farmer as an integral part of the farm organism. The gardener aims to bring together the varied cosmic forces, natural rhythms, and enlivening soil practices so that the plants can reach their full potential in flavor, fragrance, nutrition, beauty, medicinal strength, and spirit. Ideally a biodynamic farm is diversified and operates as much as possible as a closed system— meaning that the farm generates most of what it needs to function: seeds, seedlings, manure, straw, and hay. On Avena's farm we do not have enough open farmland to graze cows or sheep or to make straw. The best we can do is purchase certified organic cow manure and certified organic straw from within 60 miles of our farm.

7. Helping each plant establish a relationship to the Earth and to the larger cosmos, which in turn supports the plant's expression of its healing gifts.

The study and practice of biodynamics is a spiritual practice for me, helping me be more sensitive and attuned to the lunar and seasonal cycles, the larger cosmic rhythms, the health of the soil, the elemental beings, and a larger Spirit that guides my life. This daily awareness helps me be mindful and grateful while I engage in gardening, teaching, and herbal tasks.

ELEMENTAL BEINGS

Those who dwell among the beauties and
mysteries of the earth are never alone or weary of life.
—Rachel Carson

In his book *Cosmos, Earth and Nutrition: A Biodynamic Approach to Agriculture,* Richard Thornton Smith, a former geography professor at the University of Leeds in England, writes:

> Biodynamics addresses a system of energies underlying life processes ... when we use the word biodynamics there is a presumption that we are dealing not simply with the visible forms of nature or of agriculture, but with the underlying forces or energies. These life forces create and vitalize nature's forms—from the germinating seed to the developing mammalian embryo.

But there was something more which lay behind Steiner's motivation to speak on agriculture. He is on record as suggesting that a time would come when it would be very difficult for us to grow our crops owing to the declining vitality of the earth. We are free to interpret this as we like in relation to the world we currently inhabit, but this vision was a contributing factor to his instructions for a number of special preparations. The task of these was to strengthen connections between plant life and cosmic forces (also known as ethers) and to support the essential activities of what Steiner called "elemental beings."

Above: Tibetan prayer flags.

Opposite: Black cohosh flowers *(Cimicifuga racemosa)* with pollinating bumblebees.

Steiner spoke about the important role of the elemental beings in gardens, farms, and hedgerows in a series of lectures he gave in 1923 in Dornach, Switzerland. These lectures are currently available in the book *Harmony of the Creative Word* (formerly titled *Man as Symphony of the Creative Word*). He depicted the activity of four different groups of elemental beings and their associations with the four elements of earth, water, air, and fire. As a gifted clairvoyant, Steiner could see into the different realms of nature and described how the elemental beings, through specific tasks, serve the whole of nature. What Steiner called elemental beings are different expressions of life forces (etheric forces).

Gnomes are associated with the Earth Element and the realm of roots

Undines are associated with the Water Element and the movement of water in the stem and leafy parts of plants

Sylphs or Fairies are associated with the Air Element, the light forces, and the flowering processes of plants

Fire-spirits or Salamanders are associated with the Fire Element, and the warmth forces that help flowers transform into fruits and seeds

The first book to appear in the Western tradition describing elemental beings was written by the famous physician, scientist, and theosophist Theophrastus Bombast of Hohenheim, also known as Paracelsus. His book, *About Nymphs, Sylphs, Pygmies and Salamanders and Other Spirits,* was published in 1589. Contemporary books about nature spirits include *The Findhorn Garden, Nature Spirits and Elemental Beings,* and *Perceiving Plants: Experiencing Elemental Beings.*

In their book *Awakening to the Spirit World: The Shamanic Path of Direct Revelation,* authors Sandra Ingerman and Hank Wesselman define Shamanism as "the most ancient spiritual practice known to humankind and is the ancestor of all our modern religions." They speak about the nature spirits as "the hidden folk—the faeries and elves, the trolls and forest guardians who are present in so many myths and stories. The hidden folk remind us of a magical time in our lives before the veils between the worlds were closed to us through cultural conditioning" (p. 32).

Many indigenous peoples around the world still acknowledge the presence of nature spirits and elemental beings at work in their gardens, fields, forests, shrines, temples, sacred wells, and healing centers. We see this in places such as Nepal, Tibet, India, Bhutan, Sri Lanka, Bali, and Japan. In Scotland, Ireland, and England there are still people who perform rituals that honor the elemental beings, the natural rhythms of the sun and moon, and the sacredness of water and stones. In many Native American traditions the Little People, as well as the seasonal rhythms of the year and the cosmic influences of the sun, moon, and planets, are honored and revered.

Joe-pye-weed flowers *(Eupatorium purpureum)* and monarch butterflies.

The late Marjorie Spock, a student of Rudolf Steiner and author of *Fairy Worlds and Workers,* wrote:

> But what would become of the earth—and of ourselves—if [the fairies, Little People, elemental beings] were to vanish? Without gnomes there would be no solid ground to stand on, no firm bones to make a scaffolding inside us and the animals, no logical structure in our thinking; without undines no life-giving lakes and streams and oceans to have dry land, no green vegetation, no circulating fluids in our bodies, no liquid world of feeling in our souls; without sylphs no movement of air nor blessing of the light, no flowers, no purposeful wills; without fire-spirits no warmth, no hearth to sit by, no fruits or fragrances or grains, no fire-core of selfhood. For all this, and a great deal more, we owe the fairies thanks and recognition. Our whole world lives in soul and body because of the presence of the Little People in it.

Marjorie Spock and Rachel Carson

Marjorie Spock met Rudolf Steiner in Switzerland in 1923. After returning to the United States from Europe, Marjorie lived and gardened on Long Island, New York. In the late 1950s, she complained when the government began indiscriminate aerial spraying of DDT over wide areas of the countryside, which included her biodynamic gardens. When the spraying continued, Marjorie joined with 11 other people to bring a case against the U.S. government for its continued spraying of DDT.

In 1960, after a federal judge denied their petition, the case was taken to the U.S. Supreme Court. Marjorie began sending daily reports about the case's progression to various people, including Rachel Carson. Rachel had an interest in this case because of the research she was doing for her book *Silent Spring*. Though the Supreme Court ended up refusing to hear the case, Justice William O. Douglas dissented in that decision, stating that the concerns raised by scientists and experts regarding DDT and pesticides warranted the court's hearing the case.

Marjorie left Long Island and lived out her remaining years in Sullivan, Maine. She passed away in 2008 at the age of 103. I regret that I never spoke with Marjorie about DDT or her correspondence with Rachel. The conversations we did share centered on Anthroposophy, biodynamics, goats, herbs, and compost. My respect for both Marjorie and Rachel Carson is enormous. These important women helped set the stage for the environmental movement.

In 1962 Rachel Carson wrote:
"*Honeybees and wild bees depend heavily on such 'weeds' as goldenrod, mustard and dandelions for pollen that serves as the food of their young... Such plants are 'weeds' only to those who make a business of selling and applying chemicals.*"

In the booklet *The Agricultural Individuality: A Picture of the Human Being*, written by Martin W. Pfeiffer, is a chapter titled "Towards Caring for the Elemental Beings," in which he describes Steiner's view of the connection between earthworms and bees and the four elemental beings. Steiner associated earthworms with both the mineral earth and the moist earth and the elemental beings known as gnomes and undines. Whenever I see or gently hold an earthworm in my hand, I visualize the presence of gnomes and undines working on behalf of soil and plant life and I thank them.

In *Harmony of the Creative Word*, Lecture 7, Steiner said this:

> ... for no root could develop if it were not for what is mediated between the root and the earth realm by these remarkable root spirits, which bring the mineral element of the earth into flux in order to conduct it to the roots of plants. I am of course referring to the underlying spiritual process.

Steiner also lectured on the significance of bees and their association with the air and fire elementals: the sylphs and salamanders. He said "the insect bee is accompanied wherever it goes by the fire spirit" and "the sylphs guide the bees to the flowers." Steiner was clear that both sylphs and salamanders are needed for the flowering and fruiting processes to be complete. In the agricultural lectures Steiner stated: "In nature, and I cannot stress this often enough, all things are related to each other."

I pause more often as I watch bees, butterflies, and birds flying in and around Avena's garden and hedgerows, observing the quality of light that surrounds their wings. Various trees and shrubs grow along the edges of my garden, including a large hawthorn hedgerow we planted in 1997. Hedgerows are known to offer protection to birds, and according to Steiner, bird life is closely associated with the sylphs. I believe the planting of hedgerows or even a few trees is beneficial to our winged friends, both seen and unseen.

"In nature, and I cannot stress this often enough, all things are related to each other."
—Rudolf Steiner

Right: Female common yellowthroat warbler amidst Avena's hardy kiwi vines.

Opposite: A honeybee collecting pollen and nectar from a poppy (*Papaver somniferum*).

Jillian stirring the BD 500 prep.

BIODYNAMIC PREPARATIONS

Meditation is inseparable from the biodynamic approach. It comes with the territory. It renders us more receptive, more welcoming. It's a needed discipline as we evolve from the religious to the spiritual and move away from judgmental thoughts and reactions. Meditation is time set aside to dwell in the spirit.

—Sherry Wildfeuer

THE BIODYNAMIC FIELD SPRAYS:
Horn Manure (BD 500) and Horn Silica (BD 501)

Chanting and praying while mixing herbal preparations comes naturally to me, reinforced by the four months I spent studying in Nepal in 1980. Stirring a tiny handful of the biodynamic preparations BD 500 or BD 501 for one hour in three to four gallons of rainwater feels both ancient and familiar. Singing and repeating prayers for the well-being of all living beings while stirring the BD preps has come to be my own simple practice. There are numerous ways for people to be attentive and intentional while stirring, whether alone or in a group. Practices that temporarily stop our busy minds, inviting us to concentrate and connect with the elements and cosmic forces, are enlivening and nurturing activities to engage in.

Integrating homeopathic remedies and flower essences into my herbal work has helped me better understand the biodynamic principles behind potentizing small amounts of composted cow manure (BD 500) and pulverized silica (BD 501). These two preparations, made from different substances, are each packed into cow horns at different times of the year and buried in the ground (described below). Steiner perceived cow horns as natural vessels for receiving the cosmic forces and focusing them into the respective preparation.

BD 500 affects the biochemical processes in soil by enhancing microorganism growth in the soil, increasing earthworm activity, and improving root growth in plants. BD 501 works directly through the plant, stimulating and regulating leaf growth and supporting the flowering and fruiting process. BD 501 influences the atmosphere around young plants and the development of a strong plant structure and form. It enhances the flavor and fragrance of plants and ensures longer storage capabilities in vegetables such as carrots and potatoes. Peter Proctor, author of *Grasp the Nettle,* recommends that BD 501 be sprayed 12 months after the first application of BD 500. Peter's practice is to let the BD 500 begin to work on improving the structure of the soil for one year and then introduce the radiant silica force of the BD 501 during the next year's growing cycle.

Horn Manure (BD 500)

In Maine, April and May and September are the months we work with BD 500. The preparation is stirred and sprayed in the early evening, when the daily rhythm of the Earth is contracting and drawing downward. This makes sense as the forces of the horn manure are being directed downward to the soil and roots. Around 5 p.m. I place a small, golf-ball-size bit of BD 500, which looks like deep brown compost, into a large ceramic pot that contains three to four gallons of clean rainwater. The day chosen is based on the biodynamic calendar's guidelines (root phase, waning moon, and weather). The place I stir is a quiet spot near the garden where I can fully focus without being interrupted or distracted. With my arm I vigorously stir the water, making a vortex, then changing direction, another vortex, another change of direction, back and forth, rhythmical, meditative, sometimes chanting, sometimes praying, sometimes silent. Each vortex is stirred so that a hollow is created all the way to the bottom of the bucket before changing direction. The process of creating a vortex introduces the larger cosmic forces into the water, and the water then becomes the carrier of the life force held within the BD 500. After an hour of stirring and enlivening the BD 500, both the preparation and the one who has stirred are invigorated and transformed.

I keep three clean, five-gallon, food-grade plastic buckets and wooden-handled wallpaper brushes for the sole process of applying stirred biodynamic preparations to the garden. It's important to apply the BD 500 within two hours of stirring. Water from the ceramic pot is poured into a plastic bucket, and with brush in hand, I walk around the gardens, dipping and flinging, releasing and spreading the tiny droplets of this special soil preparation on bare soil in the garden and around the trees and hedgerows that surround the garden. And as I did when stirring, I keep my mind and heart focused on the healing task of walking and spraying the BD 500, sometimes singing, sometimes laughing, sometimes being silent.

Steiner, Pfeiffer, and other early biodynamic practitioners understood that BD 500 enhances soil fertility and earthworm activity, improves soil structure and humus formation, stimulates the soil's micro-life and beneficial bacteria, increases the moisture-absorbing capacity of the soil, and encourages good germination of seeds. Pierre Masson, author of *A Biodynamic Manual,* says the use of BD 500 also helps to regulate the acidity of the soil and stimulate the growth of root systems.

BD 500 uses fresh biodynamic or organic cow dung from a lactating cow. The dung is stuffed inside cow horns (not a bull's horn) and buried in the ground in late September and left for six to eight months. We choose a sunny area in the garden that has good-quality and well-drained soil and intuitively feels appropriate for burying the cow horns. We dig a hole that is around 18 inches deep and wide enough for the number of prepared cow horns. The horns are placed with their open ends facing downward, allowing the fresh cow manure to be in direct contact with the earth. We use a mixture of compost and good-quality garden soil to place around and over the horns, completely burying them. We then place a wooden stake in each of the corners to remind us of the exact location of the hole. Some biodynamic farmers use a dowsing rod when choosing the site to bury their cow horns.

Well-made BD 500 has a sweet earthy fragrance, the aroma of humus. It should be moist and a brownish-black color when the horns are dug up. The preparation is not of good quality if it is too dried out, moldy, green, or smells like manure. Richard Thornton Smith writes: "BD 500 represents a highly concentrated, life-giving, manuring force. [It] is the product not only of the original manure's potency, but of its intensification through infusion with the winter life forces of the earth."

We unearth the horns in June (later than biodynamic farms located in warmer climates), scoop the composted cow manure into glass jars with glass lids (old glass canning jars work great), and immediately place the jars inside a wooden box lined with peat moss. The smell of fresh cow dung is no longer present. This box is kept in a dark, cool corner of our barn away from electromagnetic waves. During the first two to four months of storage, it's a good idea to check the jars to ensure that the finished preparation has enough moisture yet isn't too soggy. If too dry, add a bit of rainwater. If too wet, use a hand tool to aerate the composted manure. BD 500 can be stored for up to three years.

As weather patterns become more erratic, I believe the biodynamic preparations will be greatly beneficial in supporting the health of plants, animals, and various

ecosystems. During times of drought, consider spraying BD 500 three times a week. As mentioned above, the BD 500 preparation enhances the soil's ability to absorb moisture, thus ensuring healthy root structures and succulent plants.

Horn Silica (BD 501)

Horn silica preparation (BD 501) is most often stirred and sprayed in the garden at dawn, preferably when dew still decorates spider webs and leaf margins, adding a magical feel to the early morning air. Steiner taught that the Earth has a daily rhythm: an out-breath with each dawn and an in-breath with each evening. With the Earth's out-breath, the fine mist of the silica, being sprayed around the garden with a backpack sprayer, gently embraces the plants and is carried upward within the structure of each plant that the silica mist surrounds.

Lindsey spraying the BD 501 at dawn.

BD 501 is applied to areas of the garden where plants have established good root systems and developed their first true leaves. I pour fresh rainwater into my ceramic crock the evening before stirring the BD 501. Between 4:30 and 5:00 the next morning (when the moon is in an ascending period; refer to the section "Biodynamic Planting Calendar"). I place one teaspoon of the pulverized silica (BD 501) into three gallons of rainwater (sprays two to three acres) and rhythmically stir for one hour in the same way the BD 500 is stirred. This potentized water is then poured into a clean backpack sprayer. I or another gardener quickly walk around the gardens, using an upward gesture with the wand, imagining the fairies' delight as the fine droplets of the silica mist envelop their realm. I aim to be spraying before the sun's first rays creep over the hillside and enter the garden. Instructors of biodynamics say it is best to spray on sunny or even misty mornings when there is little or no wind and the temperature is below 70°F.

Richard Thornton Smith says this about silica:

> [It] enables the plant to be more sensitive to all those qualities contained in the light of the day. Silica enhances the absorption of far-infrared frequencies to which water and organic molecules resonate. This is likely to confer antioxidant properties, which translate into healthier growth and better keeping quality.

Close observation of the changes in the plants' stages of growth and the weather is what guides me to spray BD 501. During a spring and summer when adequate sun and rain occur, we use BD 501 two to three times. I aim to spray the *Rosa rugosa* and hawthorn hedges and the hardy kiwi and schisandra vines in early June, when these plants are in bud but before the flowers have opened, and only if there is sufficient rainfall. The influence of the early morning light and the light forces of the BD 501 enhance the development of their flowers. If the weather has been damp and cold in May and continues into June, then I begin using the BD 501 with more frequency throughout the garden to inspirit the bud and flower formation of the well-established perennials, shrubs, and vines. Be mindful when spraying BD 501, as plants that are too young, weak, or suffering from dryness can be damaged by this preparation (biodynamic farmer Jean-David Derreumaux told me of an apple grower he knew who sprayed BD 501 in the late afternoon to avoid leaf burn). If we go through a dry, hot spell in July, then I hold off spraying the annual crops of calendula, borage, and sacred basil until early August and only after a few rainfalls have adequately moistened the soil. A quivering and expansive quality radiates from the plants receiving BD 501. It's as if a gentle rainbow has opened every single cell and they're singing.

BD 501 is made from good-quality quartz crystals that are finely pulverized to a talcum powder consistency. The horn silica's purpose is to aid the light absorption capacity of plants, strengthen the form-giving aspect that occurs as the plants grow, and ensure that their flavor, fragrance, and nutritive qualities are of very high quality.

To make BD 501, the powdered silica is mixed with rainwater, forming a paste-like substance, and then placed into cow horns. The ends of the horns are covered with soil or clay to prevent the loss of the silica paste and then buried in the ground the same way we bury the horns containing the BD 500 preparation. The time of year for the BD 501 to be held within the earth is from spring to autumn, when the sun's light forces most actively enter the earth. Silica is related to warmth and light. You can see this whenever sunlight shines through a quartz crystal hanging in a window. Burying the BD 501 in the ground when the days are longest allows the pulverized silica to more fully absorb the light forces that are strongest in summer. BD 501 is best stored in a glass jar with a loose fitting lid on an east or northeast windowsill where it receives morning light. I keep the jar on a windowsill in our barn. Every so often when I walk by the window I pick up the jar and give the silica powder a shake.

Top: Cow horns, stuffed with fresh cow manure, prior to being buried.

Bottom: Forming a ball of BD 500 (horn manure). This ball is placed into 3-4 gallons of rain water and stirred for 1 hour.

Steiner taught that the Earth has a daily rhythm, as mentioned above, and a seasonal rhythm, a continuous cycle of expansion and contraction. The period of out-breathing for the Earth's life forces is spring to autumn, and the period of in-breathing for the Earth's life forces is autumn to spring. Hillary Wright, in her book *Biodynamic Gardening*, writes beautifully about the rhythm of the seasons:

> If you think of the earth as a single living entity, then imagine its circulation system being the water that's everywhere on the planet, endlessly transforming from rain to rivers to oceans to mists. The rhythm of the seasons acts like a pulse, expanding in spring and summer then contracting in autumn and winter.

Barrel Compost

The late biodynamic researcher Maria Thun farmed in Germany from 1952 until 2012. She created a preparation that is referred to by different names: barrel compost, manure concentrate preparation, cowpat pit, and cowpat preparation. This preparation, as well as all the other biodynamic preparations, can be either made on your farm or purchased (see the Resources section at the end of the book).

A group of us have begun to make this preparation around the time of the spring equinox. We place approximately 12 gallons of fresh cow manure onto a wooden board. On top of the manure we sprinkle 4 ounces of crushed eggshells from uncooked organic eggs and 18 ounces of finely ground basalt, a volcanic rock dust. For one hour, two to three of us use spades to cut, shovel, and turn the manure over and over, walking around the pile as we "stir" the manure through the motions of shoveling and turning.

After the hour of stirring we shovel half the "dynamized" manure into a hole in the ground approximately two feet deep whose sides are lined with cement bricks. (Some people place old wooden barrels, bottoms removed, into a hole two feet deep, thus the name "barrel compost".) We place one set of the biodynamic compost preparations (see info below) into the manure, shovel the rest of the manure into the hole, insert the second set of compost preps two to three inches into the manure, and then spray the valerian compost preparation (after it has been stirred in water for ten minutes) over the manure. The hole is then covered with

a wooden board and the inoculated cow manure is left to decompose for four to six weeks. The pile needs to be aerated with a garden fork after four to six weeks. At twelve weeks, depending on weather and temperature, the barrel compost is ready to be used.

To apply barrel compost to your garden, fields, or orchard, place three small handfuls of the barrel compost into three to four gallons of rainwater and rhythmically stir for 20 minutes in the same way that BD 500 and BD 501 are stirred. Apply over the land with a brush using the same method as when applying BD 500. Some practitioners first stir BD 500 for 40 minutes, and then add the barrel compost and stir for another 20 minutes. For gardens or land that have never received any biodynamic preparations, stirring the BD 500 and the barrel compost together and applying to the land twice in the spring and then again in fall is enlivening to the soil and nourishing to the elemental beings. The Josephine Porter Institute (JPI) sells a "prepared" 500 that contains BD 500 and the six biodynamic compost preparations. JPI recommends that gardeners and farmers at the very beginning of converting to biodynamic practices use this "prepared" 500.

As with anything in biodynamics, the close observations of the gardener or farmer are essential for deeply understanding the benefits these preparations offer. The influence of the preparations can be subtle. Keep a yearly garden journal that includes the dates you apply specific biodynamic preparations and compost and record anything you notice and feel about the health and vibrancy of individual plants or animals, a particular field, forest, or place on your property or neighborhood, and the overall energetic feel of the garden and of yourself. Note the quality, taste, and vitality of your herbs, flowers, vegetables, eggs, milk, or cheese. Keep track of any pest problems. Attune to the birds, bees, and elemental beings. Record the weather. Walk your garden or farm regularly. Smell, taste, and feel your soil. Take time during and at the end of each growing season to record both practical and spiritual reflections you have noticed and felt.

THE SIX BIODYNAMIC COMPOST PREPARATIONS (BD 502–507)

The benefits of the bio-dynamic compost preparations should be made available to the largest possible areas of the entire earth, for the earth's healing.

—RUDOLF STEINER

I followed the *Stella Natura Biodynamic Planting Calendar* for 12 years before ever meeting a biodynamic farmer. I was impressed with the vitality of my herb seedlings and inspired by the articles in the biodynamic calendar and books I was reading. In the late nineties, I met a biodynamic farmer, Tom Griffin, who had farmed for several years at Camphill Village in Kimberton, Pennsylvania. Tom visited my farm and showed me how to place the biodynamic compost preparations into my compost pile. We used the wooden handle of an old shovel to make five four-inch-deep holes in a recently built compost pile. In each of these holes we placed one unit of a different compost preparation (BD 502–506). We stirred the juice of valerian flowers (BD 507) in rainwater for 10 minutes, poured some of the potentized water into each of the holes before closing them, sprayed the entire compost pile with the stirred valerian water using a wooden brush, and covered the pile with a blanket of straw.

The biodynamic compost preparations are made from six different medicinal herbs commonly used by herbalists: yarrow, chamomile, stinging nettle, oak bark, dandelion flowers, and valerian flowers. Steiner perceived that these particular members of the plant kingdom contained specific elements needed by the soil to enhance plant

growth and nutrition. Over the last 85 years, many biodynamic practitioners, including some conventionally trained soil scientists, have carefully observed and recorded their experiences using the biodynamic preparations.

> The effects of the preps were researched by the United States Department of Agriculture (USDA) in 1999. Their researchers found that use of the biodynamic compost preparations could speed the composting process, better destroy pathogens and weed seeds in the material by maintaining high temperatures longer, and change the value of the resulting compost as a fertilizer by increasing the amount of nitrate. (Wright, 2009, p. 73)

Each of the six herbs is prepared in different ways, based on Steiner's guidelines, before being placed within a freshly made or recently turned compost pile. The processes used in creating each different compost preparation are described in the books *Culture and Horticulture* by Wolf D. Storl, *Grasp the Nettle* by Peter Proctor, *Cosmos, Earth and Nutrition* by Richard Thornton Smith, and *Biodynamic Gardening* by Hillary Wright.

To better understand Steiner's spiritual and esoteric philosophies, read his books *Knowledge of the Higher Worlds* and *Goethe's World View.* Many people have found that reading and understanding Steiner's writings is most helpful when done within the support of a study group.

Steiner's *Agriculture Course* explains why the soil preparations are made using cow horns and the compost preparations with different animal sheaths.

White Yarrow (*Achillea millefolium*) BD 502

Close your eyes and imagine the delicate white composite flowers of the yarrow plant. Steiner taught that the yarrow flower is connected to the light forces and bears a relationship with the elements potassium and sulfur. Potassium is needed for good root development and for warding off root infections. "Sulfur," writes Wolf D. Storl in his book *Culture and Horticulture,* "(L. sol=sun, ferre=to carry) has the function of carrying the spiritual forces into the physical medium, a process akin to sunlight being fixed into the plant." Steiner saw each tiny white yarrow flower as a vessel for receiving beneficial cosmic forces radiating from the different planets. According to Steiner, BD 502 enhances the compost's ability to help plants attract dilute quantities of trace elements that are necessary for plant growth and to strengthen and protect plants from being attacked by insects. In Lecture Five of Steiner's *Agriculture Course* he says, "Like some sympathetic people in human society who exert an influence just by their presence, and not by what they say, yarrow's mere presence in areas where it grows abundantly is extremely beneficial" (Creeger and Gardner, 1993, p. 94).

This preparation is made by placing fresh or dried yarrow flowers into a stag's bladder and hanging it in a protected place in a tree in the summer where it is exposed to sunlight. In the fall, the bladder is carefully placed into a bottomless box in the ground at the same time the cow horns are buried, and it is left until spring. When retrieving the yarrow preparation in spring, the bladder will have decomposed and left a brownish ball of yarrow flowers whose flower structure is still visible. If the yarrow preparation is really dry, add a few teaspoons of rainwater before placing it into a glass jar that has a loose-fitting lid. We store the jars of all six of the biodynamic compost preparations in the same wooden box our BD 500 is stored in.

Chamomile (*Matricaria chamomilla*) BD 503

The chamomile preparation is seen to stabilize and regulate the nitrogen process within the compost pile, encourage the breakdown of proteins to humic plant nutrients, promote the presence of calcium and sulfur, and increase the soil's life. Steiner stated in Lecture Five of *The Agriculture Course*: "Chamomile works on calcium in addition to potash, and thereby develops what can help to ward off the harmful effects of fructification and keep the plant healthy." Steiner also said: "You will find that your manure not only has a more stable nitrogen content than other manures, but that it also has the ability to enliven the soil so that plant growth is extraordinarily stimulated" (Creeger and Gardner, 1993, p. 97).

The young chamomile flowers are harvested early in the morning when the petals are horizontal, the center flower is still greenish in color and shaped like a pyramid, and a ring of pollen-covered stamens surround the base of the center cone of the flowers.

Once picked, the flowers are dried and stored in a glass jar. In autumn the dried flowers are placed inside a fresh cow's intestines and buried in a terra cotta pot in the ground until spring. The preparation is unearthed carefully and stored in a glass jar alongside the other compost preparations.

We recently made this preparation for the first time on a biodynamic farm where friends tend a small dairy herd. The fresh cow's intestines came from one of the young male calves they had given to another family to raise for food. The farmer who heads up the dairy herd was adept at stuffing the intestines with the chamomile flowers as he has such a close connection with cows. Though I do not eat meat, I am a gardener who appreciates the meditative presence of well-loved cows on a farm and values their manure for building compost piles.

Stinging Nettle (*Urtica dioica*) BD 504

Many herbalists drink nettle tea regularly to enhance the body's overall vitality. It's one of the first spring greens we eat in Maine, cooked in soups, and the first fresh leaf tea we drink daily throughout the month of May to revitalize our life force after the long winter. Nettle tea, made with the fresh spring leaves, cleanses the body and promotes wellness and wholeness. For an herbalist, it is easy to understand that the nettle preparation enhances the compost's ability to enliven the soil and stimulate soil health.

The following statement made by Steiner during *The Agriculture Course* is one for herbalists to contemplate: "When you add this (nettle) to your manure—just like the other preparations— the effect will be to make the manure inwardly sensitive and receptive, so that it acts as if it were intelligent and does not allow decomposition to take place in the wrong way or let nitrogen escape or anything like that. This addition not only makes the manure intelligent, it also makes the soil more intelligent, so that it individualizes itself and conforms to the particular plants that you grow in it. Adding *Urtica dioica* in this form is really like an infusion of intelligence for the soil" (Creeger and Gardner, 1993, p. 99-100).

The nettle preparation is fun to make with a group of people. We harvest several pounds of the fresh nettle stalks in late September. It's important that these stalks be full of fresh leaves and flowers. In order to get nettle leaves with flowers in fall, we scythe down some of our nettles in July and place the plant material in our compost piles. This midsummer harvest allows for new nettle growth to occur so that in late September the nettle is flowering for a second time. Wearing thick gloves and long-sleeved shirts, we place large armloads of freshly cut nettle stalks into a hole approximately 15-20 inches deep that is lined with burlap. We take turns getting in the hole and tamping down the fresh nettles—with long pants and boots, of course! We then place burlap over the pounded nettles, cover it with a few inches of soil, and stake the four corners. The nettle preparation is left in the ground for one year. The following September we carefully shovel off the top layer of soil and scrape off the decomposed burlap. We then lift out the nettle, which looks like compost, and place it in a glass storage jar.

Oak Bark (*Quercus alba*) BD 505

Steiner recommended *Quercus robur* for this preparation but since this species does not grow where I live in Maine we substitute it with white oak bark. A beekeepers hive tool can be used to loosen some of the outermost layers of the bark from the trunk or large branches. The pieces of bark are pulverized into a course powder using a small grain grinder. The powder is then tightly packed into the clean skull (brain cavity) of a domestic animal, preferably from a cow or sheep. We have used a sheep's skull from one of the Maine islands where flocks of sheep live freely and die of natural causes. Steiner recommended using a skull because it is a part of the body connected with the animal's consciousness. In autumn this skull is buried in a hole at the edge of a pond or swamp where there is a flow of water.

In the spring the skull is lifted out of the water and split open, and the decomposed preparation is spooned into a glass jar. Submerging the oak bark under water throughout the winter months connects this preparation with lunar forces. The moon's more reflective or contemplative forces balance any excessive regenerative forces that may occur in the compost pile. It also regulates excessive moon forces that can cause plant diseases during long periods of rain.

In *The Agriculture Course* (translated into English by Creeger and Gardner), Steiner points out:

> One plant that contains plenty of calcium is the oak. Seventy-seven percent of its substance consists of finely distributed calcium. Oak bark, in particular, represents a kind of intermediate product between the plant and the living earth element … a large number of plant diseases can be alleviated just by means of a rational method of manuring … What is needed is that the manuring add calcium to the soil. For the

calcium to have a healing effect, it has to be calcium from something living; we cannot evade the organic realm. It won't help at all to add ordinary lime or any other calcium compound that has fallen out of the organic realm. (Creeger and Gardner, 1993, pp. 100–101)

Steiner taught that the proper amount of calcium in the soil "brings harmony to the form of the plant and does not allow excesses of rampant growth to develop."

Dandelion (*Taraxacum officinalis*) BD 506

Imagine the yellow flower of the dandelion. Then envision how ethereal the seed head looks before the star-like seeds are scattered by the wind or by the breath of a curious child. Dandelion is beloved by children, highly regarded by herbalists and earthworms, and mistakenly despised by those who seek to grow a "weed-free" lawn. Dandelion flowers are one of the first to provide valuable spring nectar to honey bees and migrating ruby-throated hummingbirds. We herbalists, hungry for something fresh and green to munch on in spring, pull on our rubber boots and with baskets, knives, digging forks, and clippers head for the garden in search of this nutritionally rich and medicinally active herb.

In *The Agriculture Course,* Steiner lectured about the dandelion flower having a relationship with silicic acid and potassium:

Silicic acid, as you know, contains silicon. In a living organism, silicon too is transmuted into an extremely important substance, one that is not currently included among the chemical elements. Silicic acid is needed to draw in the cosmic factor, and a thorough interaction must come about between the silicic acid in the plant and the potassium (not the calcium). We have to enliven the soil through manuring so that it can facilitate this interaction. Steiner continued by stating: This material can be added to the manure in the same way as before, and it will give the soil the ability to attract just as much silicic acid from the atmosphere and from the cosmos as is needed by the plants. In this way the plants become sensitive to everything at work in their environment and then be able themselves to draw in whatever else they need. (Creeger and Gardner, 1993, pp. 103-104)

This holistic and spiritual approach to ensuring that the soil is enlivened and a connection to the cosmos is facilitated is familiar to herbalists who plant and prepare their own herbal medicines.

To make the BD 506, newly opened dandelion flowers are collected early in the morning, laid out on a nonmetal screen to dry, and covered with thin muslin cloth to lessen the flowers' impulse to go to seed. After drying for two to three days, the dried flowers can be stored in a glass jar. The BD 506 preparation is made in autumn by carefully packing the slightly moistened dandelion flowers inside a bovine mesentery about four inches in diameter.[4] This dandelion "packet" is then stitched closed with thick cotton thread. The prepared packet is buried in a dry location somewhere on the farm, making sure to mark the spot so it can be found later, as it will remain in the ground until the following spring. The Earth's forces, which Steiner says are most active underground during the winter, infuse the buried dandelion with cosmic forces. "Ask the cosmic forces to enter as you would ask an old friend who comes to the door," says Jennifer Greene, a long-time anthroposophist and founder of The Water Research Institute in Blue Hill, Maine.

Valerian (*Valeriana officinalis*) BD 507

Valerian grows wild and abundantly along the coast of Maine. Avena's gardens are no exception, with volunteer valerian plants appearing in the field adjacent to our barn. In July, around sunset, we harvest a basket full of the delicate pinkish-white valerian blossoms (¼–½ pound) and immediately press them in a small herb press to obtain the fresh juice (such a small herb press is available from Horizon Herbs). Biodynamic farmer Jean-David Derreumaux describes his method for making the valerian prep:

Once I have pressed out the juice from the flower I keep it in a jar at room temperature with a lid loosely tight. It takes a bit more than a week to ferment. In that time it has decanted as well. I then pour the fermented juice in a dark amber bottle, filtering the part that has decanted. I try to size the bottle so the quantity fits in to the top so there is no air to start with. I store it with the other preps. The smell tells of the progress of the fermentation. It actually is unpleasant for some people at the beginning. The quality seems to improve over time, the smell revealing it.

The valerian preparation is spread over the top of the compost pile, unlike the other

---

4   The mesentery used is the lining of the digestive tract of a cow or ox. "The significance of the mesentery is that it is a membrane which encases and protects major internal organs which do not experience outer sensations. Such membranes are sensitive and able to store images, and therefore memories. By enclosing in the mesentery, the dandelion substance effectively becomes an organ, retaining outer impressions drawn in by the silica. The preparation therefore enables plants to be highly sensitive to their surroundings and therefore to the organism of the whole farm. A kind of alchemy is involved, for plants then benefit not only by what is in the tilled field, but also by what is in the soil of the adjacent meadow, or of the neighboring wood or forest." (Smith, 2009, p. 120)

five preparations, which are only placed within the pile. We take about one teaspoon of the liquid and place it in a clean bucket with four to six quarts of rainwater, stirring for 10 minutes and creating the same vortex as when stirring the BD 500 and 501. I first pour a few tablespoons of the BD 507 into each of the holes that contain the other five compost preparations. Then with the wooden brush I spread the "dynamized" valerian water over the entire compost heap. Biodynamic practitioners say spraying the pile with BD 507 is like placing a warm blanket over the compost heap. The valerian preparation, associated with the warmth forces, has been used by farmers to raise their soil temperature by one to two degrees. For protecting a crop from an early frost, like apple blossoms, two teaspoons of the valerian preparation can be diluted into three to five gallons of rainwater, stirred for ten minutes, and sprayed over an acre. In *The Agriculture Course* Steiner stated, "If this diluted valerian juice is applied to the manure in a very fine manner, it will stimulate the manure to relate in the right way to the substance we call phosphorus" (Creeger and Gardner, 1993, p. 104). Phosphorus has a special affinity for promoting flower seed and fruit development.

Horsetail (*Equisetum arvense*) BD 508

Decoctions or tinctures of horsetail (*Equisetum arvense* and *E. hyemale*) are used by herbalists for a variety of health conditions. Horsetail is commonly found growing in damp soils, sandy banks, and along the edges of wet fields and swamps. *E. arvense* is the species used for the BD 508 preparation and is easy to identify, as it is the only species that sends up a spore laden spear in the spring before sending up several ribbed, hollow stalks bearing whorls of reduced scale-like leaves. Early May is when I collect the young three-to-four-inch-tall horsetail plants for medicine. Some of the plants I tincture, some I infuse in organic apple cider vinegar, and the rest I dry for use in teas. Biodynamic practitioner and author Pierre Masson says horsetail can be collected and dried from June through late July, once it is more mature, for use in the BD 508 preparation.

In *The Agricultural Course*, Steiner mentioned using a decoction of horsetail to prevent or treat a variety of fungal diseases such as rust, mildew, and brown fruit rot. Biodynamic practitioner Peter Proctor says that BD 508 can be used to "reduce excessive water forces around the plant and so reduce the risk of fungal disease." During a severely rainy spring I spray the horsetail preparation onto herbs like the bee balms and the black cohoshes, which are vulnerable to mildew. Biodynamic vegetable farmers will use the BD 508 to prevent potato and tomato blight and lettuce fungus. Pierre Masson (*A Biodynamic Manual,* pp. 94–95) describes using a combined fresh or dried horsetail (¼ part) and nettle tea (3/4 part) in the greenhouse and open fields to stimulate the natural defense of plants.

Hugh Courtney writes in the introduction to the book titled *What Is Biodynamics?:*

> Using the horsetail herb can actually attract or encourage rain a very significant percentage of the time, provided that it is used when the Moon is in a watery constellation (Cancer, Scorpio, or Pisces). Similarly, if horsetail is applied when the Moon is in a fire or warmth constellation (Aries, Leo, or Sagittarius), it appears that excessive moisture conditions can be curbed. While all this is very difficult to prove, one has only to try a few times the "sequential spray" technique. Using this spray during the time the Moon is in the appropriate constellation, one can experience directly that the horsetail herb is a very powerful tool to ameliorate or balance and harmonize the watery element.

Courtney advises that all nine of the preparations "need to be used as a totality rather than in isolation from one another in order to achieve a harmony and balance of forces."

There are a few different ways to prepare the BD 508. I place one and a half ounces of dried horsetail into one gallon of rainwater and simmer for one hour. I then let it cool overnight. The next morning I strain out the herb and dilute one part decoction to nine parts rainwater. I stir it for 20 minutes in the same way as when stirring the BD 500. I then pour the dynamized tea into my backpack sprayer. Three gallons of the diluted horsetail preparation treats approximately one acre. Some practitioners make their horsetail decoction from fresh plant material, covering two pounds of the horsetail with rainwater and following the same procedure for simmering, straining, stirring, diluting, and spraying. And other practitioners let the horsetail tea ferment for two weeks before using it. Biodynamic farmer Jean-David Derreumaux told me he stores his fermented tea in the refrigerator until he needs to use it. Pierre Masson says it is best to not spray BD 508 more than 10 times on the same crop during the growing season as this may disrupt the soil's ability to build up its own fungal flora (Masson, 2011. p. 110). Dried horsetail should be used within the year it is harvested.

Horsetail is valued by herbalists for its high silica content. The spring-harvested horsetail provides easily assimilated silica and can be taken orally in tea, tincture, vinegar, or juice. Horsetail is an important medicinal herb for strengthening bones, cartilage, muscles, nails that break easily, and hair with split ends. Herbalist Matthew

Wood uses horsetail in combination with Solomon's seal root for rebuilding damaged cartilage, strengthening joints, arteries, and veins, and reducing varicosities. Matthew also uses small doses of the tincture (three drops) for people who have multiple allergies and for people who "do not have the confidence to plunge into life" (Wood, 1997, p. 254).

It's my hope that this introduction to biodynamics will pique the interest of herbalists, herb gardeners, and gardeners in general to study biodynamics further and to apply the biodynamic preparations to their gardens and compost piles. Biodynamics outlines an unusual and holistic approach to farming and gardening for helping to restore ecological and spiritual balance to our gardens, farms, and bioregions.

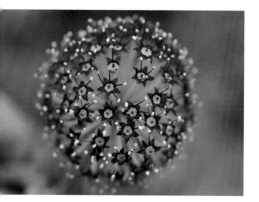

> *Biodynamic farming offers humans a path to develop our inner capacities for helping our planet to be a place for life to flourish.*

The Josephine Porter Institute (JPI) publishes information on sequential spraying of BD 500 and BD 501 for periods of drought or during periods of prolonged rain. JPI makes and sells all the biodynamic preparations, sells books and its newsletter, *Applied Biodynamics,* and offers educational programs and biodynamic consulting services. (www.jpibiodynamics.org)

FOLLOWING NATURE'S RHYTHMS

We are different people in each season with different needs for food, amount of rest, and sleep. Our energies vary: we are extrovert, introvert, far more affected by moon-rise and moon-set than we may realize. If we analyze the qualities of each season, we see that we, too, act out the drama of birth, growth, maturity, decay, and rebirth every day of our lives.

—Adele Dawson

My earliest memory of spring is sitting with my father's mother, Katherine, in an orchard and watching the apple trees bloom. I was around five. Katherine lived in a small apartment across the hall from where my family lived. Her close proximity allowed us time together for spontaneous outings. My grandmother's relationship to the natural world was unique because of her hearing impairment. While she never verbalized this to me, she mentored me by noting year after year when the first mayflowers (*Epigeae repens*) would bloom or what day the hummingbirds returned. My grandmother and I would sit on

the porch during thunderstorms, smelling the air and delighting in the colors and shapes of the lightning streaking across the sky. She knew when the wild strawberries were ripe and in her younger years had spent hours with her children collecting various berries and wild apples. My grandmother died before I reached my teenage years. Her love of the natural world, especially birds, was the legacy she left me.

I was raised in a rural town in Maine where the changes of the seasons are obvious and the night sky is free from light pollution. My father and his siblings were some of the early pioneers to hike up Mt. Washington[5] in winter with wooden skis tied to their

ABOVE: Blooming mayflower.

OPPOSITE PAGE: Ruby-throated hummingbird landing on a magnolia blossom in Avena's garden.

---

5   Mt Washington is the tallest mountain in New Hampshire and the second tallest mountain east of the Mississippi River.

canvas packs. When I was three years old my father put me on wooden skis, taught me how to lace up my boots, and sent me out to walk up and ski down a small hill near our home. The rhythm of winter, like the rhythm of spring, summer, and fall, is alive in every cell of my body, thanks to my family and to my life as a gardener.

DAILY RHYTHMS

The rising and setting sun set the daily rhythm. It's natural for humans to rise and rest according to the sun's rhythm. From an herbalist's perspective, the rising, resting, expanding, and contracting rhythms are important elements in maintaining harmony and balance in the body. According to Steiner, the Earth's daily rhythm is seen to expand in the morning near sunrise and contract in the evening near sunset. Steiner referred to the Earth's expansion in morning as a breathing out and the Earth's contraction at sunset as a relaxing in-breath. He also spoke about the seasons having a distinctive rhythm—expanding in spring and summer and contracting in fall and winter.

In the spring and summer garden, the plants' sap starts to rise during the soft light of dawn. It's my favorite time to be in the garden. In the early morning I harvest the more delicate flowers like roses, borage, heartsease pansy, and mullein, for they feel freshest and most vitally alive to me. Any gardener who is attuned to the early morning rhythm senses that the life force is strongly present in flowers and leaves at this time. Early morning is also when most vegetable farmers harvest their leafy greens. Lettuce, arugula, and basil picked at noon is wilted and lacks flavor and fragrance.[6] Dr. Rudolf Hauschka, a student of Steiner's

---

6   Steiner taught that "chaotic forces permeate the whole plant kingdom at noon and midnight" (*Nutrition*, p. 143). Not a favorable time to be harvesting or watering the garden.

and author of *Nutrition: A Holistic Approach* and *The Nature of Substance: Spirit and Matter,* spent years doing research at the Ita Wegman Clinic in Arlesheim, Switzerland. His research showed that plants harvested early in the morning contain more vital force.

Herbalists harvest roots for drying or for preparing into tinctures in the early spring or autumn, when the above-ground part of the herb is barely visible. Early in spring the energy of a plant is still focused in the root. I dig roots like marshmallow, nettle, and comfrey when the tips of their leaves first emerge out of the ground. It's different for spring-dug dandelion and yellow dock plants. For these two herbs, I wait for their fresh green leaves to develop before gathering both their roots and leaves. This is because I tincture both the roots and leaves of these herbs.

In the fall, I dig the roots of ashwagandha, astragalus, elecampane, and echinacea once the upper part of the plant has died back and their seeds have been collected. I follow the biodynamic planting calendar, harvesting roots on "root" days. If I am going to dry the roots for winter teas I wait to dig them until nearer the new moon, still on a root day, when there is less water in the roots. The possibility of roots rotting is much greater if they're harvested near the full moon when their water content is higher.

LUNAR RHYTHMS

For almost thirty years I have used the We'Moon calendar[7] to inspire and inform my understanding of how the moon's changing rhythms affect my emotional moods, weather patterns, and the garden. I am awed by the magnificence of this heavenly body and fascinated

---

7 The We'Moon calendar is an astrological lunar calendar full of beautiful art and writing by *womyn* from around the world. www.wemoon.ws.

by how the moon moves in relationship to the sun and Earth and the other planets. I am grateful to live in a rural area where I can easily see the crescent moon in the western sky at sunset, the sliver of the balsamic moon in the eastern sky at dawn, and the rising and setting of the full moon. The moon's ever changing shape reminds me to pay attention to the cycles and rhythms of the natural world and to notice how they influence my own internal rhythms.

There are four different lunar rhythms described in this section: the waxing and waning (synodic) rhythm, the sidereal rhythm, the ascending and descending rhythm, and the apogee-perigee rhythm.

The Waxing and Waning (*Synodic*) Rhythm

Perhaps the most familiar lunar rhythm to anyone who pays attention to the moon is the rhythm of expansion and contraction during each monthly lunar cycle. The moon waxes and expands from the new moon to the full moon and then slowly disappears as it wanes from full back to new. The period from new moon to new moon is a 29.5-day cycle and is known as the synodic period.

From an astrological perspective, the period from the new moon to the first quarter is a time for beginning new projects. As the moon continues to grow full, this time is for nurturing and expanding ideas and projects, and for planting seeds. Steiner mentioned in his Agriculture Course that planting seeds a few days before a full moon with rain soon to follow would allow for good germination.

The full moon, rising as the sun is setting, is a time of heightened energy, a time for ceremony or celebration with friends, and a good time to bring a project to completion. As the moon starts to wane, the third-quarter moon is a time for letting ideas settle and mature. As the moon grows smaller and appears as a tiny thumbnail in the eastern sky at dawn, this signals a time for inner reflection and quiet contemplation. The moon

---

Stella Natura calendar

The astronomical Constellations of the Zodiac are the star pictures of unequal length that we see in the sky, through which the Sun, Moon, and planets travel as they circle the Earth. These are used in the Stella Natura Calendar. They're distinct from the Signs of the Zodiac used in most popular astrology, which are 30 degree divisions of the ecliptic carrying the same names, beginning with 0 degree Aries at the point where the Sun rises at the Spring Equinox. Due to the rhythm known as the Precession of the Equinox, this point moves back 1 degree in relation to the fixed starry background every seventy-two years. The two zodiacs coincided at the time when Ptolemy was first mapping the stars, but now the constellation Pisces stands behind the Sun at the Spring Equinox, and by 2375 it will be Aquarius. Maria Thun's research with plants has consistently shown the effects of the Constellations rather than the Signs.

—from the Stella Natura calendar

disappears from the sky for approximately one day at the dark of the moon and then the waxing cycle begins again with the crescent moon appearing in the west at sunset.

The full moon by definition is in opposition to the sun which is why we see her full face each month reflecting the sun's light. Because the full moon and the sun are in opposition to one another the rhythm of the full moon will be "opposite" the rhythm of the sun. For instance, in winter, the full moon is high in the sky, following the arc of the summer's sun. In summer, the full moon is low in the sky, following the low arc of the winter's sun. Wherever you live, may the varied shapes, colors, and moods of the moon call you outside, often.

Sidereal Rhythm

The moon's sidereal rhythm refers to the time it takes the moon to complete its 27.3-day cycle through all 12 zodiacal constellations. When gazing at the moon from the Earth, the moon lies in front of one of these 12 constellations for an average of two to three days. Some of these constellations or star groupings are larger than others, which means it takes the moon longer to pass in front of larger constellations like Virgo than smaller ones like Cancer. Familiarize yourself with one of the biodynamic planting calendars to better understand this lunar sidereal rhythm.

The Ascending and Descending Rhythm

The ascending and descending rhythm (or arc) of the moon is completely different from the waxing and waning rhythm of the moon. This arc is hard to describe in words (as it appears from Earth). *The Stella Natura Calendar* prints a two-page image depicting the imaginary line called the ecliptic—the path of the sun as it passes through the zodiacal starry background during its annual journey. This image is based on the ancient astronomer Ptolemy's descriptions. Every month the moon travels this same path

moving between the highest zodiacal constellation, Taurus, down to the lowest zodiacal constellation, Scorpio, and then back up again. When the moon is ascending (moving from Scorpio to Taurus), the sap of plants move upward, leading some biodynamic practitioners to theorize that the above-ground part of the plant may be more potent during this period. When the moon is in a descending period (moving from Taurus to Scorpio), the sap moves downward, resulting in a corresponding concentration of plant energies in the root. *The Stella Natura Calendar* and Maria Thun's biodynamic calendar outline the ascending and descending arc of the moon for each month.

Depending on the weather, I aim to collect tree barks for medicine when the moon is in an ascending period. Maria Thun recommends applying the horn silica preparation, BD 501, during the ascending moon period and the horn manure preparation, BD 500, during the descending moon period. (Further discussion about the ideal time for spraying these preparations is discussed earlier in the section on biodynamic preparations.) Keep good garden notes when following these movements. Please, remember that it is more important to stir and spray the biodynamic preparations when your schedule and the weather are favorable than to become overwhelmed by all there is to know about the effects of lunar rhythms. In other words use common sense when putting the lunar rhythms into practice.

Apogee and Perigee

The moon's orbital path around the Earth is elliptical. Once a month on the *Stella Natura Biodynamic Calendar* the letters Ag and Pg will appear in column 5. Ag means the moon is apogee and is most distant from the Earth. The letters Pg mean the moon is perigee, the time when the moon is closest to the Earth. Noted on the specific day of each month on the calendar is a grey period marking the 12 hours before and after the moon is perigee, a reminder that this period of time is not favorable for planting seeds or transplanting seedlings. Some biodynamic research has shown that when the moon is perigee, seeds are more vulnerable to attack by pests and fungus, especially in wetter climates.

SEASONAL RHYTHMS

Around our planet, different cultures celebrate the changing seasons and the planting and harvesting calendar with song, dance, food, and ritual. During my second trip to Nepal in February 1992, I visited a Tibetan Buddhist family in the eastern foothills of the Himalayas during the Tibetan New Year festivities. The family made elaborate and colorful yak butter statues and placed them on their home altar, each statue having a specific meaning. On a designated day, the villagers entered the nearby Tibetan Buddhist monastery and joined the monks in chanting and ritual. The ritual included bringing a large statue of Buddha outside into the town's center and carrying it around the village while the monks and villagers chanted. It was clear that this same ritual, marking the new year and the shift towards spring, had been reenacted for hundreds of years.

In April 2008, I accompanied Ecuadorean ethnobotanist and healer Rocio Alarcon for two weeks in the Basque Country, where she was collecting herbs and traditional herbal knowledge for her Ph.D. thesis. One late Saturday afternoon, our host invited us to come along and hear her choir perform in a 1200-year-old church. The church was situated up a valley in a very small village. Much to our surprise and delight the church service ended with a group of designated people carrying a large carved statue of Mary outdoors and walking her around the village. The priest and the villagers followed, singing and celebrating spring. This ritual happens once a year and only at the beginning of the planting season.

Extraordinary

The great mystery
is that in every ordinary
person, place, thing, event
the great mystery
is present.

—JUDITH PERRY CARPENTER

Each May the staff at Avena Botanicals dances the maypole to honor spring.

For gardeners living in the northern hemisphere, the spring and fall weather clearly mark the beginning and end of the growing cycle. From early May through late July the days are long. Once August arrives, the evening light becomes noticeably less, and the night temperatures begin to drop. The midpoint between the summer solstice and the fall equinox, August 2, is called Lughnasadh by followers of the old Celtic calendar, and it ushers in the harvest season. Well-tended gardens and orchards that have been blessed with the right amount of sun and rain overflow with food, flowers, and herbs at this time of the year. The rhythm slowly shifts in August from expanding to contracting and continues contracting until the winter solstice. In November, tools are cleaned and put away, garden beds are covered with compost and straw, and harvest records collated. The days are short and the nights long. My work schedule and internal rhythm at the winter solstice look and feel very different than my summer solstice schedule. I become quiet and reflective in winter whereas in summer my energy is active and outward. The winter and summer solstices and the spring and fall equinoxes feel like the gatekeepers and guardians of the different seasons. I pause at these times to reorient my inner and outer life.

From the Earth's perspective, the sun moves higher up into the sky from midwinter through midsummer. The life forces that support plant life are stronger at this time. In midsummer another shift occurs and the sun's energy starts to wane. The arc of the sun gradually drops in the sky back to its lowest point, which occurs at the winter solstice. Wherever you live, you may find it interesting to note the place on the horizon where the sun rises at the solstices and equinoxes. At the winter solstice, the sun rises at a point south of east, and at the summer solstice it rises at a point north of east. I have come to know where the sun rises throughout the year because of my morning meditation practice. I laugh at myself during the fall and winter months for I am like an indoor cat, seeking the sun's first rays to warm me during my morning meditation. To be warmed by the sun means I have to reposition my cushion slightly each week, always keeping an eye to the place on the horizon where the sun rises.

OPPOSITE, TOP TO BOTTOM: Jewelweed blossom (*Impatiens capensis*). A favorite late summer flower of the ruby-throated hummingbirds.

Honey bee collecting pollen from an *Echinacea purpurea* flower.

Male goldfinch nibbling on Greek Mullein seeds (*Verbascum olympicus*).

Echinacea purpurea seeds.

BELOW: Late summer in Avena's gardens.

Sandy collecting St. Johnswort
(*Hypericum perforatum*) flowers for
Avena Botanicals oils and salves.

THE BIODYNAMIC PLANTING CALENDAR

The reason for using the biodynamic calendar is that one can further strengthen the plant and improve its ability to produce the desired crop. Using specific timings focuses or concentrates forces from which the different parts of plants can benefit. This improves pest resistance and yield. It also contributes to produce offering the best quality nutrition, not only in terms of conventional parameters but containing the life forces so essential in our food.

—RICHARD THORNTON SMITH

Over 50 years ago, Maria Thun, a biodynamic practitioner from Germany, began researching and recording her observations of how the germination of seeds and the growth of plants are affected by the monthly movement of the moon. Her research led to the creation of a biodynamic planting calendar.[8] Inspired by Maria Thun's research while gardening in Switzerland, Sherry Wildfeuer returned to the United States and in 1978 began publishing the *Stella Natura Calendar*.

For centuries the 12 zodiacal constellations, listed below, have been associated with one of the four primal elements: earth, water, air, and fire. Biodynamic calendars designate the most favorable times for planting root crops, leaf crops, flowers, and fruits and are based on the daily astronomical movement of the moon through these constellations. Other astronomical rhythms, including eclipses, the lunar and planetary nodes, and the position of the moon in its path around the Earth (i.e., apogee and perigee), also affect the growth of plants. These various astronomical relationships are noted on the biodynamic calendars.

- Root crops are planted when the moon moves through the constellations associated with the earth element: Taurus, Virgo, and Capricorn.

- Leaf crops are planted when the moon moves through the constellations associated with the water element: Cancer, Scorpio, and Pisces.

- Flowers are planted when the moon moves through the constellations

---

8   Refer to Chapter 6 in the book *Cosmos, Earth and Nutrition* by Richard Thornton Smith for a comprehensive description of working with astronomical rhythms and the biodynamic calendar.

associated with the air element: Gemini, Libra, and Aquarius.

- Fruit and seed crops are planted when the moon moves through the constellations associated with the fire element: Aries, Leo, and Sagittarius.

Different biodynamic planting calendars are published and available in several countries, including India, Sri Lanka, Italy, France, the United Kingdom, and the United Sates. Each focus on lunar rhythms and use standard astronomical data when outlining the movements of the sun, moon, and inner planets. Besides using the calendar for guiding the activities of sowing seeds and transplanting seedlings, many biodynamic practitioners use the calendar for pruning, grafting, weeding, beekeeping, compost making, soil cultivation, applying biodynamic preparations, and harvesting specific crops. Sherry Wildfeuer writes in the 2012 *Stella Natura Calendar*, "I would consider it tragic if this calendar were to be used merely as a utilitarian tool to improve one's crops. I hope it will inspire a sense of wonder and gratitude toward the powers of the universe which surround us and make life possible."

In order to receive the greatest benefit from following these cosmic rhythms long-term biodynamic practitioners advise that gardeners and farmers new to the calendar should at a minimum be using organic methods. This makes sense as organic and biodynamic practices support life's natural processes. As research has shown, the application of the biodynamic preparations offers another whole dimension of health and vitality to the life of the soil and crops. This having been said, even those who have been farming or gardening conventionally should not be discouraged from using the guidelines given in the calendar.

The best way to learn about planting with the rhythms outlined in the planting calendars is to begin using one. Keep good garden records. I felt a great relief when I began using the *Stella Natura Calendar*, as it assisted me in better organizing the thousands of seeds I sow each spring in my greenhouse. For example, we seed our medicinal root crops, like astragalus, ashwagandha, and elecampane, in the greenhouse in March on days listed on the calendar as root days. Sacred basil and blessed thistle seeds are planted indoors on leaf days in late April, and calendula and cosmos are planted in the greenhouse on flower days in mid- to late April. Every winter I use the previous year's seeding and outdoor transplanting schedule for guiding the creation of the new year's schedule, keeping in mind that weather plays a major role when organizing outdoor activities. Gardeners have to be flexible on a daily basis or else stress will constrict our hearts and the joy of gardening will be diminished.

The *Stella Natura* calendar hangs on the wall where the gardeners keep our notebooks and another one stays in Avena's medicine making room. Avena's gardeners and medicine maker create a weekly work schedule based on the calendar's recommendations … and weather. Weather and the needs of one's family, co-workers, and farm animals are always part of the daily dance. *Thank goodness for root days* is one of my sayings. If two days of heavy rain occur in late May on the leaf

Opposite page, top to bottom:
Early spring willow flowers.

Spilanthes acmella flowers and leaves.

Comfrey root.

Right: Teasel seeds (*Dipsacus sylvestris*).

days we planned to transplant out 300 lemon balm seedlings, then we look towards the next series of root days for transplanting our leaf crops along with transplanting root crops. Our biggest annual flower crop is calendula. We transplant 1000 calendula seedlings into the garden in mid May, another 1000 in late May, and a third round of 1000 calendula seedlings in late June. If heavy rain or intense heat prevent us from transplanting calendula during a flower time, then we look to the next series of fruit days.

Every biodynamic calendar marks the most favorable time for planting root crops, leaf crops, flowers, and fruits. The first thing I pay attention to is which constellation the moon is moving through in relationship to the type of crop I need to plant. From there I start charting out my seed planting schedule. Next, I note on the calendar the root days that occur when the moon is in a descending period in April, May, and September, as these are ideal times for spraying BD 500 or barrel compost. I then mark the flower days when the moon is in an ascending period in June and July as possible dates for spraying BD 501. This can all change due to weather, but making a plan in winter helps me organize some of the yearly biodynamic gardening tasks. I notice what phase the moon is in, waxing or waning, by watching the sky. And I celebrate the eight seasonal holidays with friends because these festivals offer spiritual inspiration and connection to the yearly cosmic rhythms.

I use the calendar as a guide. For sowing seeds in the greenhouse the calendar is fantastic. For most of our transplanting and root harvests we follow the calendar as best as circumstances allow. It's not always possible to follow the calendar when harvesting short-lived annual flowers like calendula, chamomile, and elder flowers, or daily leaf crops like lettuce, arugula, and parsley. We harvest medicinal flowers for Avena Botanical's elixirs and extracts in the early morning when the blossoms are newly opened and their life forces feel most vital. We wait until the dew has disappeared before gathering flowers to dry, ideally when the moon is in an air or fire sign. Once dry, these flowers are blended into teas and infused into medicinal oils and salves.

We adhere to the biodynamic calendar's grey areas—commonly called black-out periods—which indicate unfavorable times to perform gardening activities. We use the

black-out periods as an opportunity to work on gardening-related activities such as catching up with paperwork, sharpening tools, cleaning the herb-drying room, and collecting cardboard at a nearby recycling center (used for sheet mulching paths and garden edges). The grey periods are clearly marked on biodynamic calendars. Biodynamic practitioners and researchers have come to recognize that there are particular times during each lunar cycle when the lunar forces are out of balance. These grey periods include eclipses; moon at perigee; the time when the moon, Mercury, and Venus cross the ecliptic (nodes); and when the moon transitions from one zodiacal constellation into another. Becoming familiar with the symbols for the constellations, the planets, the nodes, and the course of the moon will help you more quickly understand the daily and monthly planetary movements outlined in the calendars. Each calendar lists the symbols so you can keep referring to them as you are learning.

In his series of agricultural lectures, Steiner mentions that plant growth is affected by cosmic forces that radiate from the moon and planets. In Lecture 6 he states that "along with the moon's light the entire reflected cosmos comes towards the earth." Studying Steiner's work can help one comprehend these concepts more fully, but it is equally helpful for each individual to become a student of the heavens by paying close attention to the daily and monthly movement of the moon and planets. Go outside regularly and admire the night sky. Learn to identify the constellations and the visible planets. Find a friend to share meaningful new moon and full moon excursions with.

Being introduced to biodynamic agriculture through the calendar was a significant turning point in my life. Time spent with my grandmother, the gift of Juliette de Bairacli Levy's herb book, my introduction to Buddhism in Nepal, and my study and practice of biodynamics have all guided my life as a gardener and herbalist. An awareness of the daily, weekly, and seasonal rhythms along with using the biodynamic preparations, meditating, and studying spiritual texts are practices that help me to be mindful and joyful. I care deeply for the well-being of the plants, the pollinators, and the people I serve and am grateful for the opportunity to study and practice biodynamics and herbalism as a way to contribute something positive and healing to our world.

Our time is calling for us as farmers and gardeners to wake up to our potential to be active inwardly as well as outwardly. In this way our work can bring the spirit of healing into the very food (and herbs) that is grown through our care.

—SHERRY WILDFEUER

OPPOSITE: Lindsey transplanting calendula seedlings.

III

Living in Harmony with the Seasons

INTRODUCTION

By imbibing nature's foods in harmony with the seasons we are able to foster enormous strength within the body and develop a kinship with the seasonal rhythms. This is why the ancients remained attentive to the cyclical changes of both the inner and outer rhythms during each season and its transition into another. Food also feeds the mind as well as it nourishes the body; once the body is properly nourished, the mind is fed. Fulfilled with essential nourishment, we can easily transcend challenges to reclaiming our consciousness. Once harvested, consciousness influences how we respond to each and every circumstance and experience in our lives. Health, healing, and harmony are the ultimate result of nature's essential nourishment.

—Maya Tiwari

My life as a gardener and herbalist is intricately intertwined with the cyclical changes of the seasons and the moon. I am always listening for the return of the male wood cocks in late March for they're a sign that early spring has arrived. Soon after their return I begin planting seeds in my greenhouse. Once the full moon in late May or early June has passed I begin transplanting the tender annuals such as sacred basil and nasturtiums into the garden. Though an early June frost can surprise us on the coast of Maine, usually we are without frosts from June until the full moon in September or early October. Observing the waxing and waning of the moon and the cyclical changes of the seasons keeps me attuned to nature's rhythms.

I was introduced to the idea of eating with the seasons when I was a student living in Nepal. Soon after returning from Nepal I sought out acupuncture treatments to support my health and to keep me connected with a traditional system of medicine that is organized around the seasons and the elements. Traditional Chinese Medicine's (TCM) descriptions of the seasons resonate with me, which is why I included them here.

Besides the spring, summer, fall, and winter seasons that most Westerners are familiar with, TCM recognizes a fifth season: late summer. Each of the five seasons has specific associations with different emotions, organ systems, foods, flavors, climates, elements, and activities. Learning the energetic qualities and the foods associated with the different seasons brings balance and harmony into one's inner and outer life. Our overall health, happiness, and well-being are deeply rooted in the rhythms of the natural world. However, it is up to each of us to observe how we feel during the different seasons and not rely solely on ancient systems of interpretation. When we're aware of and aligned with the cycles of Nature, we have the opportunity to enhance our vitality and quality of life, feel connected with and supported by the Earth's life-affirming forces, and live in a truly sustainable and awakened way that respects, nurtures, and honors all life.

SPRING

In the northern hemisphere, especially where I live in Maine, the light of day noticeably shifts at the beginning of February. February 2 is the mid-point between the winter solstice and the spring equinox. In northern Europe this time is called Imbolc or Candlemas. In

Ireland this day is known as Brigit's Day. On a recent trip to Sicily, I happened upon an old church celebrating February 2 with an elaborate candle-lighting ceremony. I sat near the back of the church behind an old woman who kept herself distant from the rest of the church's members. I followed her up to the main altar when the priest began handing out long thin candles. Once back in our seats, the old woman lit her candle with a cigarette lighter, then silently turned and lit mine. These ancient rituals follow a seasonal calendar where no words are needed.

A stirring of energy occurs in both the mind and body when February arrives and the feeling of deep winter starts to shift, even though in places like northern Maine, Montana, and Labrador the cold and snow still persist. When the ice of winter begins to melt in March and April, the water element of winter gives way to the wood element of spring. If you have rested deeply in winter and feel replenished, the awakening of your vital life force in early spring will naturally begin slowly and without hesitation. Sap buckets hanging off spigots on the sugar maple trees are a sign that the energy of spring is beginning. Drinking a cup of the clear, pure maple sap before we boil it into syrup is one of my favorite spring tonics. The days become significantly longer after the spring equinox, March 20–21, and with this light comes the call to rise earlier and engage in more physical activities. Spring is the season of renewing the body's vitality by using spring greens and bitter root teas, fresh juices and lighter foods, engaging in more outdoor activities, and clearing out our homes, closets, cars, and stressful internal affairs.

In TCM, spring is associated with the liver, gallbladder, and the element of wood. TCM says the liver is responsible for many physiological processes, including the smooth flow and storage of blood, digestive functions, flexible tendons and ligaments, and balanced emotional states. When the liver is congested, feelings of anger, frustration, and irritability arise. Women may experience more mood swings, heightened feeling states, and irregular menstrual cycles when their liver's health is out of balance.

Acupuncturists guide their clients to enter early spring by envisioning what they want to do in the year ahead. A balanced uprising of energy in spring will support us through the rest of the year in accomplishing our visions, dreams, and creative projects. Reducing or eliminating stimulants like coffee, recreational drugs, and alcohol will help people who constantly feel agitated or who awake in the night between 1:00 a.m. and 3:00 a.m. Stimulants aggravate the liver, which in turn disrupts sleep, digestion, and the ability to make clear decisions.

People worldwide have traditionally called upon spring greens to cleanse the body and clear the mind. Because the taste of sour resonates with the liver, spring greens tend to be bitter, sour, and pungent in flavor and are perfect for improving liver health. My favorite edible spring greens include dandelion, nettle, watercress, lamb's quarters, violet, chickweed, chicory, amaranth, red mustard, and arugula. Fresh nettle is my favorite leaf tea, and burdock root, dandelion root, and yellow dock root are my favorite springtime tinctures. The roots of astragalus, codonopsis, and fresh ginger continue to be added to my soups and teas for immune and digestive support as spring weather in Maine can be cold, damp, and rainy.

The strong winds of spring often contribute to a variety of springtime ailments such as stiffness, tension, headaches, allergies, and skin conditions. Many people, eager to feel the sun on their skin, dress too lightly and end up with a cold or stiff neck. Scarves, hats, and jackets help protect us from the erratic and windy weather of spring.

Getting into the garden and planting seeds is a satisfying mid-spring activity. When there's a gentle spring rain, go outside and breathe in its reviving essence. If you're planting perennials or trees, envision what they'll look like in five or ten years. A healthy liver supports our ability to have long-term visions, and the gall bladder helps sort through the decisions needed to manifest those visions. A balanced wood element also helps our ligaments and tendons to remain flexible and our eyes moist and clear. Flexibility in the body aids flexibility in the mind.

Late Spring

The mid-point between the spring equinox and the summer solstice, May 1, known as May Day or Beltane, finds me more in the garden than in the greenhouse. There are fewer seeds left to plant indoors. Garden beds are in need of weeding, and the list of gardening tasks suddenly becomes enormous. Gardeners can become easily overwhelmed when the dynamic, thrusting energy of spring arrives in full force. Daily

SOUR TASTING FOODS AND SUPPORTIVE HERBS FOR SPRING

Vegetables and grains: Barley, quinoa, rye, arugula, lightly sauteed Asian greens, dandelion greens, mesclun salads, wild mustard greens, steamed nettle greens, violet leaf and other wild edible greens, beets, carrots, cabbage, fermented vegetables, umeboshi plum paste

Miso: Adzuki, brown rice, dandelion and leek, sweet white

Sea vegetables: Arame, dulse, kelp, nori, wakame

Culinary herbs: Chives, cilantro, garlic, ginger, lovage, parsley, rosemary, thyme

Medicinal herbs: Burdock roots, birch leaf, chickweed, dandelion roots and greens, nettle leaf, yellow dock roots and greens, sweet cicely leaf, violet leaf and flower

Wood's colors: green

The following foods are best avoided throughout the year: Iced-cold food and drinks, artificial sweeteners, canned soup, excessive amounts of coffee and salt, frozen processed TV dinners, hot dogs, pretzels, potato chips, and stimulants. These foods when eaten regularly weaken the body's digestive, endocrine, and immune systems.

meditation and a written list of the week's work help me stay grounded and more at ease in my mind and body.

May is when we spray the biodynamic soil preparation called BD 500, complete our spring root and nettle leaf harvests, and begin transplanting several thousand seedlings into the garden. There's a tremendous push for us farmers and gardeners to plant our annual crops before the heat of summer has us weeding, watering, and harvesting. My acupuncture friends remind me to watch each seasonal shift closely, as transition times can be stressful to the body and mind. Eating nourishing food, pausing to notice my breath throughout the day, and taking warm evening baths are practices that help me feel more mindful and present amidst a demanding work schedule.

SUMMER

Summer in Maine feels short, sweet, intense, and outward. The heat and expansive nature of summer is with us in northern New England from June through early August. Though the arrival of the summer solstice, June 21, marks the beginning of summer, for us in northern New England it denotes that the sun has reached its most northern point in the wheel of the year and that the next six weeks are meant to be lived as actively outside as possible.

TCM associates summer with the fire element, the heart and small intestine, and the emotion of joy. The heat and long days of summer foster a feeling of optimism, which can lead

to a feeling of ease and harmony in one's body and relationships. The dynamic push of spring is over and we are entering into the fullness of summer. Schools are on vacation and the fun of picnics, swimming, hiking, picking berries, and watching plants bloom has begun.

Summer is the season for creating beautiful meals from the abundance of locally grown vegetables and fruits. Eating lighter and easy-to-digest foods such as fresh salads, sprouted legumes, nuts, seeds, and locally crafted cheeses and drinking cooling

flower and leaf teas (lemon balm and mint) connect us to nature's bounty. It's important to balance the intense summer heat with watery and cooling foods such as cucumbers, strawberries, watermelons, and quinoa. Too much raw food, extremely spicy or fried foods, iced drinks and ice cream, and red meat and alcohol weaken the digestive tract and interfere with digestion, assimilation, and elimination. The function of the small intestine in Chinese medicine is to govern ingested fluids and foods, sort and separate the pure from the impure, send food to the large intestine for reabsorption and elimination, and send impurities to the bladder. We support the small intestine by choosing to eat a combination of cooked and raw fruits and vegetables.

It's essential during the heat of summer to balance and ground outer activity with meaningful conversations with friends and with moments of reflective alone time. The summer's tendency to be overly outward and active inhibits the heart's fire from feeling truly warmed and satisfied. The heart needs experiences that nourish a steady flame. If we burn ourselves out in summer and head into the harvest season depleted, then the body and mind are unable to sort through what is valuable and what needs to be discarded; the tendency to spiral into negative thought patterns, excessive thinking, and depression is more likely to occur.

When the heart is settled and peaceful, a person is warm, friendly, humble, clear minded, open hearted, and happy. Problem solving takes less effort as the heart and mind are clear. According to TCM, the heart regulates blood circulation and sweat, supports mental activities, controls consciousness, and houses the *Shen*. In her book *Wood Becomes Water: Chinese Medicine in Everyday Life* , acupuncturist Gail Reichstein writes:

> Called shen in Chinese medicine, the spirit has a number of different manifestations. Because of this, many practitioners find it helpful to distinguish between a "Big Shen" and a "little shen." The "Big Shen" refers to Spirit in general and is an expression of the force of creation. It embodies peace, a sense of belonging, and a knowledge of the unity of all things. Little shen, on the other hand, refers to those aspects of spirit that are directly connected to the Heart and the Fire element. Whereas Big Shen involves the whole being and all the elements, little shen describes what the Heart gives: warmth, love, a sparkle in the eyes, and vitality in the body. It also refers to what we call the mind in Western culture, which is responsible for consciousness, awareness, and thought. Little shen is more easily disturbed by minor imbalances of the Heart and can cause symptoms like anxiety, disturbed sleep, or mania; Big Shen is disturbed by severe trauma and manifests its disturbances in serious mental illnesses. (pp. 80-81)

Playing, laughing, singing, praying, meditating with friends, and spending contemplative time in nature are activities that nourish the heart and the *Shen*. Taking a few drops of a rose petal elixir directly on the tongue and soaking in a bath full of floating rose petals will also calm and nourish *Shen*.

Late Summer
The mid-point between the summer solstice and the fall equinox occurs on

BITTER AND COOLING FOODS AND SUPPORTIVE HERBS FOR SUMMER

Vegetables and grains: Amaranth, quinoa, lightly sauteed Asian greens, mesclun salads, cooked kale and broccoli, celery, romaine lettuce, scallions, turnips

Fruit: Local fruits, including blueberries, strawberries, cherries, peaches

Miso: Chick pea, millet, sweet white

Sea vegetables: Dulse, kelp, nori

Culinary herbs: Basil, cilantro, dill, parsley, nasturtium flowers and leaves, rosemary, watercress

Medicinal herbs: Blessed thistle, chamomile, hawthorn flowers, lemon balm, motherwort, rose petals, rosemary, sacred basil, yarrow

Fire's colors: bright orange, red, peach, pink

Top to Bottom:
Ruby-throated hummingbird.

Goldenrod with honeybee.

Statue of the beloved deity Tara. With love and compassion Tara relieves the suffering of all sentient beings.

August 2. Northern European peoples call this time Lammas or Lughnasadh. At 4:30 a.m. the light is dimmer, signifying a seasonal shift. I know late summer has begun when I feel a bit melancholy. The male hummingbirds start to leave the garden for their long journey south, and fewer plants show off their fragrant flowers. I pause often to watch the juvenile hummingbirds chase each other from flower to flower—they will not head south until mid-September. I am like an elder, drinking in the garden before I, too, am gone.

Years ago I attended a Buddhist meditation retreat led by Sharon Salzberg, author of many books, including *Lovingkindness: The Revolutionary Art of Happiness.* For several days we silently practiced Metta meditation, repeating short phrases that offer beneficial and healing qualities to oneself and to others. The word *metta* comes from Pali, the language of the original Buddhist texts, and means "lovingkindness," or "friendship." Late summer is an exceptionally good time to introduce or reintroduce a heart-centered *metta* practice into your life. There are many phrases you can create and repeat every day, such as these:

> *May I dwell in the heart.*
> *May I be free from suffering.*
> *May I be healed.*
> *May I know the joy of my own true nature.*
> *May I be at peace.*

In northern climates, August is when the harvests of grains, grapes, blueberries, and early apples begin. People who celebrate Lughnasadh use the last of their flour to bake special cakes, trusting that the harvest will be abundant. Late summer is associated with the Earth element. Acupuncturist Elizabeth Garber says that "ripeness and roundness" begin to transpire in the garden in late summer. As the Earth draws her energy inward, the ripening of sweet fruits and vegetables and winter squashes quickens.

The spleen and stomach are the organs associated with late summer, and they govern digestion and assimilation. The stomach takes in and digests food, and the spleen takes what's been digested and transports it throughout the body so all cells can be nourished. When we feel deeply nourished and connected to the rhythms of nature, a feeling of balance, no matter what internal or external changes are occurring, is possible. Late summer is the time to fully take in and assimilate all the experiences we have tended and nurtured since spring. The sweetness of Earth beckons us to fill our hearts and minds with

SWEET TASTING FOODS AND SUPPORTIVE HERBS FOR LATE SUMMER

Vegetables and grains: Sweet corn, barley, brown rice, millet, carrots, summer squashes, shallots, onions, sweet potatoes, artichokes, cabbage, cooked leafy greens

Fruit: Blueberries, cantaloupe, early apples and pears, figs

Miso: Chick pea, millet, sweet white

Honey

Sea vegetables: Arame, dulse, kelp

Culinary herbs: Basil, cardamom, cinnamon, coriander seed, fennel seed, garlic, marjoram, nutmeg

Medicinal herbs: Anise hyssop, astragalus root, codonopsis root, eleuthero root (Siberian ginseng), fresh grated ginger, goldenrod, licorice root, marshmallow root, oat seed, Solomon's seal root

Earth's colors: orange, yellow, brown

gratitude and to extend compassion and nurturance to ourselves and others.

AUTUMN

Autumn is a busy time for gardeners. As the days shorten and the colors and flowers of the garden fade, this is the season for fermenting and canning vegetables and preparing winter squashes and root crops for winter storage. In the herb garden, we work around the wet and cold days, digging roots and collecting fruits, berries, and seeds. The red schisandra berries start to ripen when their medicine begins to be needed. The hardy kiwi vines, covering one of Avena's arbors, offer their delicious

Top: A golden-crowned kinglet.

Middle: Hardy kiwi fruit.

Bottom: Gretchen collecting hawthorn berries.

fruits for nibbling fresh and for cooking into jams. The hawthorn hedge that surrounds Avena's garden is laden with red berries, and it's one of our most important harvests. We use a long apple pruning pole to cut the various branches that hang heavy with the hawthorns' fruits. Once the berries are on the ground, we collect them into baskets (being careful to not get poked by the thick thorns that line the branches), tincture them fresh, and lay them on screens in the drying room. The medicinal properties of schisandra and hawthorn are discussed in Chapter V.

The season of autumn brings lots of changes. The "winds of change" can blow erratically, heralding in different types of weather: sometimes dry, sometimes wet, sometimes hot, and sometimes cold. For some people, autumn feels refreshing; for others, the erratic nature of autumn is unsettling. I have observed over the years that fall is one of the hardest seasonal transitions for lots of people, both physically and emotionally.

From the fall equinox (September 21) and onward, the decrease in sunlight affects people's spirits in different ways. Fall is a time of letting go and of evaluating the projects we have tended and completed throughout the summer. It can be a time when people feel dissatisfied and judgmental towards themselves and others, especially if they're not content with what's occurring in their lives. Autumn offers the opportunity for clear and thoughtful reevaluation and the releasing of what no longer serves us, both internally and externally. Letting go allows room for what is truly precious to be present in our hearts and minds.

The lungs and colon are the organs associated with autumn. Healthy and strong lungs are vital for supplying the body with oxygen and releasing the "used" air that is made up of carbon dioxide and toxins. One of the concepts of TCM that I like is the image of the lungs extracting "clean energy" from the air, combining it with "food energy" from the food we eat and together forming the "upright energy" needed to nourish the body. The weaker and more congested the lungs are from poor-quality foods, overactivity, and exposure to toxic pollutants and allergens, the more likely a person is to experience fatigue, illness, and dis-ease. In TCM, the lungs are paired with the colon. The colon's health is vital to maintaining overall health and resiliency. The colon and the skin both eliminate waste from the body. Constipation and skin eruptions are signs that the body's eliminative channels are congested and need support.

Early autumn is when I encourage people to start making soup broths with astragalus root, burdock root, codonopsis root, and reishi and shiitake mushrooms; I also encourage teas with astragalus root, codonopsis root, cinnamon, ginger, and licorice root. These soup broths and teas (available through Avena Botanicals) strengthen the lungs and immune system. Children and adults who experience recurring colds, flu, ear and sinus infections, and pneumonia during the winter months will benefit by introducing preventive herbal teas and soups in the fall and continuing them into spring.

Pungent is the flavor associated with autumn and the time of year when we add more pungent and warming herbs and cooked foods back into our diets. Culinary herbs such as anise seed, black pepper, long pepper, chili pepper, caraway seeds, cardamom, cinnamon, clove, dill, garlic, ginger, and horseradish are beneficial, along with foods such as adzuki beans, arugula, broccoli and broccoli raab, carrots, cauliflower, cooked apples and pears, Chinese cabbage, celery, grapes, kale, leeks, mustard greens, olives, and onions.

ABOVE: Carefully pulling the hawthorn berries off their stems in preparation for tincturing.

Fall is the hardest seasonal transition for many gardeners, moving from the warmth and expansiveness of summer into the cooler and more contracted time of autumn and winter. Grief is the emotion associated with autumn. It's the time of year I see people starting to struggle with colds, sinus and bronchial infections, runny noses, asthma, allergies, wet coughs, and constipation. These symptoms are clues that the lungs, colon, and immune system need tending. Dr. Elson Haas writes in his book *Staying Healthy with the Seasons*:

Illness represents an invitation to change. I see this deeper understanding as very important. Of course, this does not deny the existence of microorganisms or accidents, but it gives us a different focus. Creating a healthy body and being in harmony with nature and one's own inner guidance will help prevent infections and other mishaps.

In the fall rhythm of the year we're called to slow down, turn inward, eat more warm and cooked foods, dress in layers, pace our work lives differently, and release old patterns, behaviors, and things that no longer nourish us. This releasing offers clarity for what serves us and how we can serve others. The ancient pagan holiday Samhain, celebrated on October 31 as the Celtic new year, is a favorite of mine. This date clearly marks the end of the growing season in the northern regions. It's when the Earth's energy is moving inward, giving me permission as a gardener to do the same.

PUNGENT AND WARMING FOODS AND SUPPORTIVE HERBS FOR AUTUMN

Vegetables and grains: Arugula, beets, brown rice, buckwheat, broccoli and broccoli raab, carrots, cauliflower, Chinese cabbage, celery, celeriac root, collards, daikon radish, turnips, kale, fennel bulbs, leeks, mustard greens, onions, potatoes and yams, winter squashes .

Fruit: Cooked apples and pears, grapes, hardy kiwi fruits

Miso: Adzuki, brown rice, chick pea, dandelion and leek, millet

Sea vegetables: Dulse, hiziki, kelp, wakame

Culinary herbs: Anise seed, black pepper, long pepper, black mustard, chili pepper, caraway seeds, cardamom, cinnamon, clove, coriander seed, cumin seed, dill, fennel seed, garlic, ginger, horseradish, marjoram, rosemary, thyme, turmeric

Medicinal herbs: Astragalus root, burdock root, codonopsis root, elderberries, eleuthero root, garlic, ginger, hawthorn berries, hyssop, licorice root, mullein leaf and flower, reishi mushroom, rosemary, schisandra berries, shiitake mushrooms, thyme

Metal's color: white

Herbalism is a cooperative healing venture among humanity, plants, and the Earth.

—DAVID HOFFMAN

WINTER

Winter in Maine can sometimes feel unending, barren, and bleak. There are days when I long to smell the delicate fragrance of the evening primrose flowers, feel the pond's refreshing coolness against my bare skin, watch a hummingbird dart about the garden, or listen to the wood thrush call the evening in. In my younger years, by late February I would feel restless for the return of spring. But as I grow older, I savor winter as the time for dreaming, for deep rest, and for renewing my body, mind, and spirit. I have become monk-like, protective of this season, using my day time for quiet contemplation and writing, my night time for meditation and dreaming. I long for weeks of uninterrupted space, where it's only the wind, the trees, and the rising and setting moon and sun with whom I engage in dialogue and prayer.

Winter is associated with the water element and the kidneys, adrenals, and bladder. These organs regulate fluid metabolism and support the endocrine system, reproductive system, bone health, and auditory abilities. TCM states that the kidneys store our "original yuan/original chi." Acupuncturist Connie Evans says that this original chi is formed by our parents at conception and is what motivates the vitality that we're born with. TCM teaches that the health of the kidneys and adrenals determines the vitality and length of our lives and governs the storage of the body's life force in the bones.

Winter is a time for being especially mindful of our energy reserves. It's when we need to draw our energy inward and refill the deep well that contains our chi. If we move from

winter into spring feeling rested and revitalized, then summer will feel energetic and expansive, and autumn will feel abundant and satisfying. If we've been unable to rest sufficiently in winter and be nourished by food, family, community, and safe and meaningful work, then over time we begin to lack the chi needed to feel healthy and vital.

Water is the most yin[9] of all the elements and needs to be contained. One indicator of a person's overall vitality is how often they arise in the night to urinate. The stronger your chi, the better the bladder contains and holds urine through the night—unless you drank several cups of tea near bedtime. If a person regularly rises several times a night to urinate, then ongoing support for the kidneys, adrenals, and bladder with food, herbs, acupuncture, and lifestyle choices can be beneficial.

TCM states that when the kidney energy is low, regardless of the season, the emotion of fear, which is associated with the kidneys, may become predominant. Besides the common health issues discussed above that affect kidney and adrenal health, there are many social, political, and environmental reasons why people, especially women and

---

9   Acupuncturist Connie Evans defines yin as a tendency towards stillness, quiet, coolness, moistness, darkness, and nurturing.

SALTY WINTER FOODS AND SUPPORTIVE HERBS FOR NORTHERN CLIMATES

Vegetables: Arugula, beets, broccoli, burdock root, carrots, celeriac root, daikon radish, fermented vegetables, garlic, kale, radish leaf, rutabagas, shiitake mushrooms; sunflower sprouts and other sprout mixes make a great addition to a grated carrot, beet, and celeriac root salad

Fermented vegetables: sauerkraut, kim chi, and other fermented vegetables (refer to the book *Wild Fermentation* for great recipes)

Grains: Barley, buckwheat, black and red rice, basmati rice, short grain rice, wild rice

Beans: Adzuki beans, black beans, black soybeans, kidney beans, mung beans, pinto beans, and yummy spiced dals made with different types of lentils

Ghee: (clarified butter) is easy to make and is warming and building to the body. 1 lb organic, unsalted butter makes 1 glass pint jar of ghee. I cook with ghee all year long.

Fruit: Blackberries, blueberries, raspberries. We freeze 50 to 70 pounds of organic blackberries and blueberries and enjoy them in cooked cereal, pies, and crisps throughout the winter. Whatever raspberries don't get eaten fresh from our patch go into pint-sized freezer bags for special winter treats.

children, live with moderate to extreme fear, which leads to emotional and physical exhaustion. The continued violence women experience daily around the world is horrific. Every nine seconds a woman is beaten or raped. Poverty, homelessness, war, lack of access to clean water and fresh food, and exposure to toxic chemicals are on the long list of devastating events that degrade human dignity and destroy the biological processes that support life on Earth. As these issues become more directly and compassionately addressed by communities everywhere, I pray that a larger shift in human consciousness will occur.

People who enter winter feeling tired and depleted are more susceptible to frequent colds and flu, low back pain, cold hands and feet, lowered sexual energy, digestive challenges, depression, joint pain, and poor memory. Dress with plenty of clothing, especially warm hats, scarves, socks, sweaters, and coats that keep the head, neck, waist, ankles, and lower back covered. The waist area encompasses the kidneys. Exposure to extreme cold or wet weather weakens the kidneys and leads to lowered immunity. Cover the waist, neck, ankles and head to prevent loss of heat. When possible, create a winter rhythm that has a slower pace and includes more rest, warm and nourishing foods, meditation and reflection, outdoor exercise, self-massage with good-quality oils, and warm (not overly hot) baths, showers, or saunas.

Rest, reflection, and contemplation are supportive practices to engage in during the winter months. Winter's rhythm begs us to slow down and retreat inward in order to renew our strength and spirit.

Miso: Darker varieties for winter

Sea vegetables: Dulse, hijiki, kelp, and wakame

Nuts and seeds: Brazil nuts, chestnuts, sesame seeds, sunflower seeds, tahini, gomasio (sesame seed and sea salt condiment)

Culinary herbs: Black pepper, long pepper, chili pepper, cardamom, cinnamon, clove, coriander seed, cumin seed, curry spices, dill seed, garlic, ginger, hingvastaka and trikatu (traditional Ayurvedic combinations), rosemary, thyme, turmeric. Most aromatic herbs are great for adding into winter foods and teas as they tend to be warming and dispersing.

Medicinal herbs: Ashwagandha, astragalus root, bee balm flowers, calendula, codonopsis root, echinacea root, elderberries, fresh ginger root, hyssop, nettle, oats, rosemary, sacred basil, schisandra berries, thyme, usnea lichen; root teas such as astragalus, codonopsis, burdock, ginger, licorice, eleuthero, and cinnamon bark. I love adding cardamom pods and cloves to Avena's Immune Tonic tea (contains astragalus, codonopsis, cooked rehmannia, cinnamon, ginger, and licorice) for a delicious winter herbal chai.

Water's colors: purple, black, blue

IV

Energetic and Elemental Associations
of Plant Parts

INTRODUCTION

Plants were described by botanists of antiquity as fourfold in nature. It's not difficult to sense the earth element, solid and dense, teeming with microscopic life, manifest in the roots of plants. Water flows through the stems and leaves of every plant, pulled up by osmosis and transpired back to the atmosphere as the exhaled breath of water vapor. The element of air is conveyed in the flowers of the plant world, blooming and communicating in the soft summer wind. And the fire element is consolidated in the fruit and seed of certain plants since fruit and seed, those inward seeds of fire, are both ripened and cured by the heat of the sun.

—Wendy Johnson

Wendy Johnson's book *Gardening at the Dragon's Gate* is one of my favorites. While reading this book I kept remembering my visit with farmer and author Masanabu Fukuoka at his farm in Japan in 1982. The spiritual practices of both Wendy and the late Mr. Fukuoka are not separate from their gardening and farming activities. In her book Wendy includes the following basic meditation and everyday life instruction given to her years ago by Zen priest Katagiri Roshi. These instructions are posted on the wall near my desk.

> *Settle your self on your self—*
> *and let the flower*
> *of your life force bloom.*

Flowering plants and trees and the cyclical movement of the moon and the seasons never cease to amaze me. I find myself standing more often in awe of what is happening inside Avena's herb garden. Walking slowly in the garden, feeling my feet touching the earth settles my heart and mind and reminds me that this moment, right now, is the most important moment there is, and that it too will change as everything around me expands and contracts, is born and falls away.

The four elements—earth, air, fire, and water—ground me in my work as a biodynamic gardener and herbalist. Each morning, whether it is summer or winter, I light incense and honor these elements. They're so fundamental to life. I also honor the element of ether. Though not included in the section that follows, the ether element represents spiritual energy. In the ancient system of Ayurveda, originally from India, it is said that ether joined with air to bring movement and form into existence.

The American spiritual teacher Adyashanti writes in his book *The Way of Liberation*:

> We should come to know that there is more Reality and sacredness in a blade of grass than in all of our thoughts and ideas about Reality. When we perceive from an undivided consciousness, we will find the sacred in every expression of life. We will find it in our teacup, in the fall breeze, in the brushing of our teeth, in each and every moment of living and dying. Therefore, we must leave the entire collection of conditioned thought behind and let ourselves be led by the inner thread of silence and intuitive awareness, beyond where all paths end, to that place of sacredness where we go innocently or not at all, not once but continually.

The Four Plant Parts: *elements, energetics, and the zodiac*

Roots are associated with the earth element.
Energetic quality of roots: groundedness, rootedness, connectedness
Constellations associated with the earth sign: Taurus, Virgo, Capricorn

Leaves are associated with the water element.
Energetic quality of leaves: movement, transformation, freedom
Constellations associated with the water sign: Cancer, Scorpio, Pisces

Flowers are associated with the air element.
Energetic quality of flowers: light, expansiveness, beauty
Constellations associated with the air sign: Gemini, Libra, Aquarius

Fruits and seeds are associated with the fire element.
Energetic quality of fruits and seeds: manifestation, fruition, carriers of memory
Constellations associated with the fire sign: Aries, Leo, Sagittarius

Please remember that what is written here about the energetic associations of plant parts is offered as a stepping stone for your own personal exploration. It's best not to rely solely on another person's interpretation. Your direct experiences with plants, animals, and weather will be most beneficial. Books such as *Plant Spirit Healing* by Pam Montgomery and *Awakening to the Spirit World: The Shamanic Path of Direct Revelation* may be helpful to those new to communicating with plants, animals, and nature spirits.

ROOTS

Vitality comes from our capacity for connectedness and presence in the world, forming a relationship to life that is rooted in compassion and generosity.

—CHRISTINA BALDWIN

Roots. Taproot. Fibrous root. Prop root. Rootlet. Rhizoid. Tuberoid. Tubercule. Radicant.

Root: that portion of the plant axis lacking nodes and leaves and usually found below ground.

Astragalus root. Ashwagandha root. Burdock root. Codonopsis root. Comfrey root. Dandelion root. Echinacea root. Elecampane root. Marshmallow root. Nettle root. Pleurisy root. Solomon's seal root. Teasel root. Valerian root. Yellow dock root.

These root names roll off my tongue like a favorite camp song from long ago. Spring after spring and fall after fall these roots are dug, shaken, washed, chopped, tinctured, dried, and sung over. Some come out of the ground quite easily, while others require more patience and physical coaxing. Some have been growing for several years, while others, for only one.

The primary function of a root is to anchor the plant in the soil, absorb water, transport water, and store food and nutrients for the plant's future use. Botanists consider plants to have six basic parts: roots, stems, leaves, flowers, fruits, and seeds. They all have unique shapes, colors, textures, and tastes. From root to seed and seed to root, a continuous movement of expansion and contraction is occurring.

Over a thousand years ago, astronomers named the 12 constellations of the zodiac through which the sun, moon, and planets pass. Astronomers matched each of these constellations with one of the four elements. Over 50 years ago, German biodynamic farmer Maria Thun began systematically planting seeds of various crops when the moon was traveling through the different constellations. She documented her results and developed a calendar system based on the moon's sidereal cycle (see page 83). For example, Maria's research showed that when the moon passed in front of a constellation associated with the earth element (Taurus, Virgo, Capricorn), the root growth of plants

seeded at this time were enhanced. Maria Thun's research is described in her book *Results from the Biodynamic Sowing and Planting Calendar*. While Maria's work is said by some to not always be replicable, I have found her work to be informative and useful for organizing my seed planting and transplanting schedules and root-digging days.

The energetic qualities of the four elements help herbalists and gardeners more deeply understand the patterns we see in plants, people, and illnesses. Herbalists often recommend the use of teas and tinctures made from roots for the fall, winter, and spring months because roots offer strength and nourishment during the colder and darker seasons. Medicinal roots and root vegetables are beneficial to those who feel ungrounded, disconnected, scattered, weak, or discouraged and those who have experienced a trauma or loss. Roots lend support to those seeking to strengthen their inner core, to stand with confidence, and to be present and rooted throughout life's varied journeys. Infinite possibilities emerge from strong roots.

OPPOSITE: Teasel roots (*Dipsacus sylvestris*)

LEFT, TOP TO BOTTOM:
Elecampane roots (*Inula helenium*)

Hand holding teasel root (*Dipsacus sylvestris*)

Chinese red root sage (*Salvia miltiorrhiza*)

MARIA THUN found that when the moon passed in front of a constellation associated with the earth element, root growth was enhanced. The zodiac constellations Taurus, Virgo, and Capricorn are associated with the earth element.

LEAVES

In music,
in a flower,
in a leaf,
in an act of kindness . . .
I see
what people
call God
in all
these things.
—Pablo Casals

Simple leaves. Compound leaves. Palmate leaves. Pinnate leaves. Perfoliate leaves. Trifoliate leaves. Palmately lobed. Pinnately lobed. Alternate leaves. Opposite leaves. Basal leaves. Whorled leaves.

Leaves: *the usually expanded, photosynthetic organs of a plant.*

Basil. Boneset. Borage. Catmint. Gotu kola. Ground ivy. Lemon balm. Lemon verbena. Nasturtium. Nettle. Oregano. Rosemary. Rose geranium. Sweet cicely. Sweet marjoram. Thyme. Leaf parts, leaf bases, leaf apices, leaf venations. The vastness of leaf terminology can be overwhelming yet invites the gardener's mind to be curious and attentive.

The scientific name for boneset is *Eupatorium perfoliatum.* "Perfoliate" describes *leaves with the margins entirely surrounding the stem, so that the stem appears to pass through the leaf.*

The shapes, colors, textures, and fragrances of leaves have stories to tell, just as do the people who come to walk in my garden. An 80-year-old man once told me that boneset was collected annually by his mother, hung upside down in bunches to dry, and given as tea to the children for winter colds and flus. He remembered the distinctive shape of the leaves and the extremely bitter taste of the tea. Boneset is so bitter that I taste the bitterness in my mouth when harvesting the flowering plant—even

before placing a piece of the leaf in my mouth. Native to northern New England and Canada, this plant grows near streams and ponds and in damp areas of a garden and reseeds itself when happy. Boneset is my favorite remedy for reducing the achiness that often accompanies a flu and for adding into a formula for healing broken bones.

The water activity happening in leaves and stems contributes to the various chemical changes occurring in a plant, photosynthesis being one of these amazingly complex chemical processes. Leaves produce food through the process of photosynthesis, and this food is transported throughout the stems and leaf structures. Richard Thornton Smith, author of *Cosmos, Earth and Nutrition: A Biodynamic Approach to Agriculture,* writes:

Water is the crucial link between the "scientific" explanation and a more metaphysical one . . . water is the great catalyst for the working of life forces. When seeds are placed in moist soil they undergo an awakening process. The seed with its cosmic imprint unites itself with the earth and the life forces it finds there, a fact reflected by the radicle being the first vegetative growth to emerge. *(p. 167)*

Leaves. Water. Movement. Transformation. Watch the phases of the moon from new to full and full to new, month after month and season after season. The moon teaches about impermanence, as it is always changing. It never stays the same. Just like the garden. Just like the gardener.

Top: Magical dew drops.

Middle: Monarch caterpillar munching on a milkweed leaf.

Bottom: Flowering boneset.

Lady's mantle plant covered with early morning dew

MARIA THUN found that when the moon passed in front of a constellation associated with the water element, leaf growth was enhanced. The zodiac constellations Cancer, Scorpio, and Pisces are associated with the water element.

FLOWERS

The world of flowers with its extraordinary repertoire of vibrant color, beautiful intricate patterns, and alluring scent brings more than a touch of paradise to our earthly lives. It also brings the possibility of healing on both physical and non-physical levels of existence.

—ANNE MCINTYRE

Aster. Bee balm. Bleeding heart. Calendula. California poppy. Cardinal flower. Clary sage. Cleome. Cosmos. Echinacea. Evening primrose. Foxgloves. Lavender. Lily. Lungwort. Magnolia. Milkweed. Nasturium. Passionflower. Pearly everlasting. Peony. Red clover. Self-heal. Sunflower. Violet. Wood betony.

Flowers: the reproductive portion of the plant, consisting of stamens, pistils, or both, and usually including a perianth of sepals or both sepals and petals.

Fresh and dried flowers prepared into teas, tinctures, oils, salves, baths, steams, elixirs, and flower essences help the body, mind, and spirit heal. Throughout the ages flowers have been included in ceremonies and rituals and have been given as gifts to honor particular passages in a person's life. The beauty, fragrances, and expansive nature of flowers offer humans the possibility of moving from feeling separate and isolated into being present in this very moment with an expanded spiritual consciousness.

My earliest memory of being awed by the fragrance of a flower is when my grandmother brought me for the first time to collect mayflowers. She was skillful at knowing where to find their tiny white blossoms, hidden under shiny green leaves that wove their way upward through leaf litter and forest floor debris. The mayflower, also known as trailing arbutus, was her favorite wildflower. Years after my grandmother passed,

I began searching for what the old timers call a *mayflower bank* in the woods behind Avena's farm. With the eyes of a child and my grandmother's gentle guiding spirit, I found a small patch of mayflowers. Visiting this bank has become a spring ritual for me like it had been for my grandmother.

Inflorescences: *the flowering part of a plant; a flower cluster; the arrangement of the flowers on the flowering axis.* The botanical names of flower shapes, petals, and sepals are numerous. A *campanulate* flower is bell shaped. A *gibbous* flower is swollen or enlarged on one side. *Papilionaceous* means butterfly-like and refers to the irregular petals (*corolla*) of a sweet pea that include a banner petal, two wing petals, and a keel petal. A *tubular flower* has the form of a tube or cylinder. Using a hand lens to look closely at flowers and tasting their pollen and nectar is what sparked my fascination and interest in studying floral morphology and pollinators.

In their book *The Forgotten Pollinators,* biologist Gary Paul Nabhan and entomologist Stephen L. Buchmann outline a newer paradigm for studying the diversity and complexity of plant/pollinator interactions proposed by ecologist Judith Bronstein. Her paradigm goes beyond the basic pollinator syndrome paradigm where biologists infer who the animal pollinators are for specific flowers based on the distinctive characteristics of the flower and its pollinators.

Judith's proposal "requires more empirical observations of pollinator foraging and flowering patterns of all the plants in a particular landscape, and better approximates the diversity of plant/pollinator interactions" (Buchmann and Nabhan, 1996, p. 67).

Judith directs gardeners and scientists alike to observe the specific floral traits important to the pollinators who visit them as well as to study the larger landscape patterns, noting the variety of flowers in a landscape and

the diversity of pollinators feeding upon them. Judith's paradigm is expanding the old model to include an awareness of the interconnectedness of all life forms.

Throughout the summer I collect and dry various flowering herbs for mixing together into winter teas: orange calendula flowers, red and purple bee balm blossoms, rose petals, white pearly everlasting buds, tiny purple catmint blossoms, the delicate lady's mantle flowers, sweet-tasting red clover blossoms, and the deep purple anise-hyssop flowers. Their colors, aromas, flavors, and songs invite me to simply be present as I blend them into teas.

ABOVE: Drying calendula flowers.

RIGHT: A nasturtium flower, a spicy edible flower.

OPPOSITE PAGE, TOP TO BOTTOM: Evening primrose flowers (*Oenothera biennis*). A favorite of the ruby-throated hummingbirds.

Pleurisy, also known as Butterfly weed (*Asclepias tuberosa*). Frequently visited by honeybees and monarch butterflies.

A joyful sunflower.

MARIA THUN found that when the moon passed in front of a constellation associated with the air element, flower growth was enhanced. The zodiac constellations Gemini, Libra, and Aquarius are associated with the air element.

In the Garden by M. C. Richards

In the garden the wild mind buds and swells

with words that cannot rest, cannot be spelled,

that turn this pink, this lemon yellow,

the color speechless saying,

"Here here the language lives intense,

love lifts its hues like sentences that sculpt the soil."

Oh how to say it,

how to bend with the words' weight yet feel them fly

in brushstrokes purple and mint and magenta and blue

completing a phrase.

 Veils of color pulse in the mind,

become syllables and tones, a music that dissolves,

makes us stutter into trills and birdsongs,

turns us loose, wilderness at play.

The player is our soul in its garden,

where words bud in the bright mouths of spring.

(*Opening Our Moral Eye*, p. 156)

FRUITS AND SEEDS

The outer work will never be puny if the inner work is great.

—Meister Eckhart

Fruit. Nuts. Nutlets. Legumes. Drupelets. Drupe. Acorns. Hips. Berries. Burs. Pepo. Pome. What wondrous names for these miraculous fruits.

Fruit: *a ripened ovary and any other structures that are attached and ripen with it.* Ovary: *the expanded basal portion of the pistil that contains the ovules; the immature fruit.* Seed: *a ripened ovule.*

Blackberries. Blueberries. Elderberries. Goji berries. Hawthorn berries.

Lingonberries. Raspberries. Rose hips. Strawberries. Schisandra berries. Apples. Currants. Grapes. Hardy kiwi. Peaches. Pears. Plums. Sea buckthorn.

All these types of fruit, and more, grow in my home state of Maine. Their ripening tells me what month and what season we are in: early summer, late summer, early fall, late fall. Their perennial nature teaches about commitment and devotion.

We planted 48 hawthorn trees in 1997 and collected our first fruits in 2006. The female and male hardy kiwi vines planted in 2000 yielded their first green fruits in 2007. Schisandra vines planted in the spring of 2000 bore their first berries in 2008. Our elderberry shrubs began producing fruit three years after being planted. The old farmers knew what it took to commit to and care for a piece of land—the soil, the trees, the animals, the birds, the barn, and each other.

Some fruits need to be planted near different varieties of the same fruit for cross-pollination. Some need more water than others. All need pollinators and soil that is rich in microbial activity. Some fruits we call food. Some we call medicine. The line between food and medicine becomes thin when we study

ABOVE: Rose hips.

OPPOSITE PAGE: Gretchen collecting hawthorn berries from Avena's hedgerow.

133

the nutritional benefits and the medicinal qualities of fruits and seeds.

Burdock seed. Coriander seed. Cumin seed. Dill seed. Echinacea seed. Fennel seed. Flax seed. Love-in-the mist seed. Milk thistle seed. Pumpkin seed. Sesame seed. Sunflower seed. Some seeds we eat, some we tincture, and some we make into tea. Some we collect, some we cook with, and some we replant.

We live in a time when the genetic modification of seeds and the dangerous mentality that drives the GMO industry need to change dramatically in order to wake up and value life, not dollars. As consumers, each of us can choose what seed companies[10] and farmers we support, what markets we purchase our food from, and how we involve ourselves in local, statewide, and national legislation that affects agriculture, farmers, pollinators, and all of life.[11]

Seeds contain the hope and genetic coding for a biologically diverse future. Each seed we hold in our hand has stories to tell of how it came to fruition, how the health of the soil, watershed, and pollinators contributed to its maturation, and who the gardeners' ancestors were, their voices guiding the next generation in the rhythmical work of planting, tending, harvesting, praying, and believing in the miracle present in seeds.

Top to bottom:

Hawthorn berries.

Ashwagandha seeds.

Blessed thistle seeds.

Opposite page: Common milkweed seeds.

10   Seed companies are listed in back of the book.

11   Author and world-renowned environmental thinker and activist Vandana Shiva continues to write, speak, and act on behalf of non-GMO seeds, organic and sustainable agriculture, and biodiversity. Her books are so worth reading. May they find their way into every high school and college classroom!

MARIA THUN found that when the moon passed in front of a constellation associated with the fire element, fruit and seed growth was enhanced. The zodiac constellations Aries, Leo, and Sagittarius are associated with the fire element.

V

*Growing, Harvesting,
and Using Medicinal Herbs*

INTRODUCTION

Attune to food as medicine, medicine as plants, plants as emanations of the Great Spirit. Let us respect our landscape by greening it up for the benefit of all beings: rocks, microorganisms, insects, plants, trees, fishes, mammals, birds and humans. May the elements of soil, sun, air, and water nourish our plantings and contribute to the balance of nature. If you find yourself separate from this balance—if your mind is fettered by the distractions of modern times, if your breath comes quick and shallow, if your body aches from stasis, if your emotions spill out of control and bring you down, then I have only one piece of advice. Go outside.

—Richo Cech, *Horizon Herbs*

Roots. Leaves. Flowers. Fragrance. Fruits. Seeds. Spirit. We humans have much to learn from the world of plants and trees and the spirit that pulses within each of them. We become whole, over and over again, by consciously attuning to these magnificent and intelligent forms of life and by gratefully ingesting plants as food and medicine. The simple tasks of making tea and cooking and eating food become holy when approached mindfully and with gratitude.

Plants support the healing process in a myriad of ways. On a daily basis we can drink nourishing herbal teas and enliven our food with the aromas, flavors, and digestive support herbs offer. These nourishing teas and culinary herbs add minerals and vitamins to our diet, aid the body's biochemical processes, and impart a special sparkle to our day. When we stop and directly experience the physical nourishment and spiritual presence of plants, we are experiencing the Earth as a living, breathing, conscious being.

ABOVE: *Rosa rugosa* bud.

OPPOSITE: Deb Soule standing at the entrance to her garden.

The establishment of respectful relationships with plants and trees is essential for the development of a gardener and herbalist. Learning about the healing properties of plants from books is beneficial. However, sitting quietly with a plant and closely observing it through its different life cycles; touching, tasting, and smelling fresh plants; and being in the environment where the herbs are naturally growing enhance a person's understanding of the health benefits and sacred qualities of plants.

Being quietly in nature, whether for a few hours or for a short time in the morning or evening, begins to shape an individual's inner spiritual understanding of the natural world in ways that far surpass book learning. This direct way of learning encourages a deeper connection with nature and with one's inner experience of stillness. From this place of stillness arises respect, curiosity, compassion, and gratitude for one's self and for all living beings.

Create simple daily rituals like bowing to trees, tenderly touching a leaf, or smelling a flower. Spend time sitting quietly with a plant or tree. Notice how you feel. Over time expand your vision to include the wider ecological community you are sitting with. Communing with plants regularly deepens and expands our connection with the green world and helps each of us feel whole and integrated. Our restored essence is then able to radiate love and joy out into the world. Tasting, touching, and tenderly holding plants are wondrous and sacred experiences. Whether it be in the wild, in a garden, in the kitchen, or in a healing bath, plants help mend and heal the broken and disconnected places within humans. There is no substitute for a direct experience with a living plant.

Top: Avena's lavender hedgerow.
Middle: Chamomile flowers.
Bottom: Red clover blossoms.

Wherever I have traveled in the world, I have always found gardens to visit. Though I appreciate and enjoy visiting botanical gardens, I am most interested in walking in the smaller gardens tended by women. It's here that I have seen and learned about the daily lives of women and the plants they cultivate and use. Usually women do not have time to spend hours alone in faraway fields and forests collecting medicinal herbs. They grow, collect, and prepare remedies from their household gardens for their families and neighbors. I have observed women in their gardens while traveling in Nepal, India, the Basque Country, Sicily, England, Ireland, Canada, and the United States. It's in my travels that I have come to realize that the healing language of plants is universal.

I hope this book will end up in the kitchens and potting sheds of women and men who are eager to plant, harvest, and prepare herbal remedies for their families and communities. I also hope that more herb gardeners and farmers will begin using the biodynamic preparations as medicines for the Earth. There is a healing intelligence in these preparations, like in herbs, that a gardener will come to understand with regular use and thoughtful reflection. May we all continuously reevaluate our lifestyles and join with others in creating sustainable and compassionate solutions to the enormous environmental and spiritual changes we are experiencing at this time.

Deb Soule in Sicily, enjoying the scent of lemon verbena.

MATERIA MEDICA

A garden, small or large, like a nuclear or extended family, gives us a protected place to grow—not only to grow herbs in a way compatible with nature, but to grow in our own psychic awareness, to cultivate our potential for being sensitive and responsible citizens of our planet and grateful caretakers of our inherited treasures.

—ADELE DAWSON

There are as many ways to describe the medicinal uses of plants as there are herbalists. I chose to include some of my favorite herbs in this chapter. They're all herbs I grow and have used for myself and with clients over many years. I organized them into categories based on the plant part used: root, leaf, flower, and fruit and seed. This organizational system stems from my biodynamic background. Remember as you study the actions of herbs and the body systems they have affinities with that this is only one way to understand their medicinal benefits. Plants have an intelligence far greater than words can express. Let your direct experiences with plants guide you in understanding their healing gifts.

This book is both practical and poetic, for that is the way of most gardeners. May herbalists and gardeners find what is written here beneficial for healing themselves and assisting others. Choosing to use herbs that are grown without chemicals, and harvested and prepared with gratitude and love, is a bold and life-affirming step on behalf of our planet. May simple herbal remedies, delicious spices, and nourishing teas once again find their rightful place in every household and community clinic.

Study rhythm, rhythm is the carrier of life.

—RUDOLF STEINER

ABOVE: Dandelion flower with honeybee.

OPPOSITE: Avena Botanicals apothecary shelves filled with herbal tinctures.

143

ASHWAGANDHA

Latin name: *Withania somnifera*

Common names: ashwagandha, winter cherry

Family: Solanaceae

Place of origin: India, Pakistan, Sri Lanka

Part used: root

Energy: warm

Flavor: bitter, sweet

Cultivating: Grown as an annual in New England and as a perennial in Asia. Start seeds indoors in late March or early April in a 70–75° F greenhouse, on a warm windowsill, or on a heat mat. Seeds need light and warmth to germinate. Carefully press the seeds into organic potting soil. Avoid covering the seeds with soil and keep them moist by watering with a gentle spray nozzle or mister. In Avena's greenhouse we use 3-inch-deep pots when planting ashwagandha seeds and place plastic domes[12] atop the pots to keep the seeds moist. Seeds germinate in 7–14 days. Seed quality and adequate heat enhance germination. You can also purchase seedlings from mail order companies (see list of Seed and Plant Sources pp. 223).

We transplant individual seedlings into 4-inch pots when they're 1–2 inches in height. Once the plants reach 4 inches in height, and there is no longer any danger of frost, we transplant the seedlings, 8 inches apart, into a hot sunny location in soil that is well drained. We mulch the seedlings with straw and rarely water them after they're initially watered. They thrive without much water. Ashwagandha is transplanted into a different garden bed over three different summers before circling back around to a bed it was previously grown in.

---

12   A note on plastic: I have been reusing the same plastic pots, plug trays, and plastic domes for over 15 years. We dip them each spring into a large bin that contains a few gallons of clean water and a diluted capful of a non-toxic sanitizing product called Sanidate. This mixture helps sterilize the pots and trays. We also use two different sizes of wooden cedar trays (available through FEDCO) when planting seeds. These cedar trays will last for years if stored in a dry place when not in use.

Collecting: In mid-October we collect the seeds of ashwagandha (ideally on a fruit day, following the biodynamic calendar) when their outer coat is red. We dry them in brown paper bags and store them in bags or glass jars. Based on the weather and the biodynamic calendar, we dig the first-year ashwagandha roots on root days using a small border fork.

We lay the roots on a long wire mesh table that is outside and wash them using warm water and a lead-free hose. The hose is hooked up to an outdoor spigot that has access to hot and cold water. We wash the roots several times before chopping them and laying them out to dry in one of Avena's herb drying rooms. The smell of good-quality ashwagandha roots is strong. Because of this, we make sure that by mid-October one of the drying rooms is empty and ready to receive only the ashwagandha roots. Once the roots are dry they can be made into tea, tincture, or ground into a powder as needed. Dried roots have a shelf life of two years when stored in glass jars in a cool, dark cupboard. Powdered herbs have a much shorter shelf life.[13] When purchasing a powdered herb, be sure to ask the date it was powdered and store in a glass air-tight container in a cool, dark cupboard.

Actions: adaptogen, anti-inflammatory, antioxidant, anti-tumor, mild astringent, nervine, immune amphoteric, rejuvenative, reproductive tonic

---

13 Some herbalists say that powdered herbs have a shelf life of about 6 months when stored properly. Storage recommendations include keeping your herbs in a tightly sealed, glass jar with a label of the herb's name and date it was harvested or powdered. Jars are best stored in a cool, dark cupboard or closet. Do not store herbs in a refrigerator, freezer or near a heat source. Whole dried leaves and flowers have a shelf life of 12 months. Dried roots, barks, and seeds have a shelf life of 18-24 months.

INDICATIONS

Gynecology: An excellent postpartum energy restorative tonic and a stimulant for milk production. Tones uterine muscles. Can be used as part of a formula for women with dysmenorrhea or amenorrhea or for women who have miscarried. Ashwagandha is one of my favorite herbs for easing a variety of perimenopausal and menopausal symptoms, including insomnia, anxiety, agitation, restlessness, fatigue, cloudy thinking, low libido, and muscle pain. It offers much benefit to women.

Immunity: Ashwagandha is highly revered in Ayurvedic medicine as a rejuvenative tonic. It's beneficial to people with autoimmune conditions such as rheumatoid arthritis, multiple sclerosis, systemic lupus erythematosus, osteoarthritis, cancer, and chronic connective tissue disorders. The root appears to have both immunosuppressive and immune tonic qualities, helping to bring a person who is either depleted and exhausted or overly stimulated back into balance (Pole, 2006, p. 133). Donald Yance, author of *Herbal Medicine, Healing and Cancer,* recommends using ashwagandha during and after radiation therapy as he says it "significantly increases white blood-cell count, reducing leucopenia caused by radiation and immune suppressive drugs" (Yance, 1999, p. 127).

Nerves: Calming and strengthening to the nervous system without being overstimulating. Effective for nervous and depleted conditions such as anxiety, nervous exhaustion, impaired memory, insomnia, ADHD, stress-induced ulcers, and chronic debilitation that are the result of physical and mental overwork. Calms the mind and promotes a deep, restful sleep.

Reproductive: Female and male reproductive tonic. Increases sperm motility and sperm count. Promotes conception.

Tissues: Improves poor muscle tone. Beneficial to children, women, and the elderly who are anemic and run down. Rich in iron—use the powdered roots in warm milk with blackstrap molasses for women with iron-deficient anemia. Beneficial for people who feel debilitated and/or who have low body weight. Helpful for reducing inflammation and edema and for strengthening muscle tone following an injury or surgery.

Thyroid: Useful in hypothyroid disorders. Enhances endocrine function.

PREPARATION AND DOSAGE

Powder: ½–1 teaspoon powder in 4–8 oz of warm dairy (or oat, nut, or seed) milk with honey or blackstrap molasses, once or twice per day. Mix the powdered roots into honey or ghee if you do not drink milk. Ayurvedic practitioners recommend using ashwagandha powder in a reproductive tissue building carrier such as milk, ghee, almond milk, or honey (Pole, 2006, p. 134).

Ashwagandha powder can be mixed in a 50/50 ratio with organic ghee and taken in teaspoon doses, 2–3 times per day.

Tincture: (1:3 or 1:5) Dried roots only, take ¼–1 teaspoon, 3 times per day.

External uses: Herbalist and Ayurvedic practitioner Anne McIntyre recommends ashwagandha roots infused in organic sesame oil as a massage oil for painful joints, frozen shoulder, sciatica, muscle spasms, numbness, and back pain.

Safety considerations: Not recommended during acute infections or for persons with high ama (thick-coated tongue and chronic constipation are signs of ama—see glossary). Because of the root's warming qualities, ashwagandha may be too heating of an herb for people whose body temperature runs hot. Traditionally used in India in small amounts, less than 3 grams/day, during pregnancy to calm a restless fetus and strengthen the mother (Winston and Kuhn, 2001, p. 39). In the West, its use during pregnancy is not recommended. Herbalist and medical doctor Aviva Romm writes: "The evidence contradicting its use during pregnancy is limited and questionable, and Ayurvedic practitioners have used it traditionally during pregnancy" (Romm, 2010, p. 202). It's best to consult with your midwife if you're pregnant, and avoid during the first trimester.

Ashwagandha may potentiate action of barbiturates (Winston and Maimes, 2007, p. 141). Avoid using if you have an allergy or sensitivity to plants in the nightshade family (tomatoes, potatoes, eggplant). I recommend that people with chronic constipation and thickly coated tongues consult with an herbalist or Ayurvedic practitioner before using ashwagandha. Constipation and/or a coated tongue are indicators that the health of the digestive system needs to be addressed before using tonifying herbs like ashwagandha. Ashwagandha is known to stimulate the thyroid and therefore not indicated for people whose thyroid is overactive.

Long-term safety: safe

Ashwagandha Milk with Rose Petal Elixir and Cardamom

Herbalist and author Anne McIntyre has been coming to teach Ayurveda at Avena Botanicals for many years. Our favorite bedtime drink is one cup of warm almond milk with ½ tsp of ashwagandha powder, a dropper-full of Avena's Rose Petal Elixir, a pinch of cardamom (enhances a meditative state), and honey. Deliciously relaxing. Anne says taking ashwagandha consistently for a year will give you the strength of a horse for the next 10 years.

ASTRAGALUS

Latin name: *Astragalus membranaceus*

Common name: milk vetch

Family: Fabaceae

Place of origin: China

Part used: root (4-year-old)

Energy: slightly warm, moist

Flavor: sweet

Perennial: hardy to minus 15°F

Cultivating: The small, kidney-shaped seeds have a hard outer coat and will germinate more effectively if rubbed with sandpaper, soaked overnight in water, or seeded into plant pots or trays that are placed on heat mats. In Avena's greenhouse we scarify the seeds with sandpaper before sowing them into pots in late March. The seeds are light dependent, so gently tamping them into organic potting mix works best. Germination generally occurs in 5–10 days. Be aware of mice: they love to nibble astragalus sprouts. We transplant the seedlings into 4-inch pots when they're 2 inches tall and then plant them into the garden in June. The plants are spaced 10 inches apart in rows. Each garden bed contains 2 to 3 rows, spaced 2 feet apart. Astragalus prefers well-drained soil that is not high in nitrogen. It's an excellent herb to plant as part of a 4-year crop rotation for improving the quality of the soil. Being a member of the pea family, astragalus helps to fix nitrogen. We grow astragalus on a south-facing slope that has excellent drainage and in an area we rarely water. Be sure to not over-water or over-fertilize this wonderful herb.

Collecting: In mid-October, prior to digging the roots of the astragalus plants, we collect the seeds before the pod's papery outer layer opens. The seeds are dried in paper bags. We dig the 4-year-old roots (on a root day following the biodynamic calendar) in mid- to late October. The long, yellowish tap roots can grow to be 3 to 4 feet long. We use a transplanting spade (longer and thinner than a standard spade) for digging medicinal roots, as it's important to dig a deep, steep-sided hole and obtain as much of the whole root as possible. The outer bark of good-quality astragalus is thin, and the roots feel firm and a bit flexible. Once our roots are washed and chopped, we lay some out to dry and

the rest are decocted fresh in water for 12 hours before being ground and made into a tincture.

Actions: adaptogen, antioxidant, qi tonic, diuretic, hepatoprotective, immune tonic, lung tonic

INDICATIONS

Cardiovascular: Improves cardiac blood flow and physical stamina. Astragalus can be used to protect the heart against damage from toxins and viruses (McIntyre and Boudin, 2012, p. 27).

Digestion: Strengthens digestion and improves absorption of nutrients. Anne McIntyre recommends using astragalus when peristalsis action is slowed down because of weak bowel muscles. In Traditional Chinese Medicine (TCM), astragalus is considered to be a spleen qi and can be used for conditions such as lack of appetite and fatigue (Winston and Maimes, 2007, p. 148).

Gynecology: Aviva Romm discusses the use of astragalus in a treatment protocol for women with HPV (human papillomavirus) and chronic cervicitis. She recommends an external herbal treatment for genital warts alongside the oral use of a tincture containing astragalus root, ashwagandha root, reishi mushroom, echinacea root, and usnea lichen for boosting the immune system (Romm, 2010, p. 269). Astragalus is an excellent herb for lifting prolapsed organs—uterus, rectum, bladder, hemorrhoids—and for raising the energy of women who feel physically and mentally exhausted. It's a safe and valuable herb for women to use for postpartum support, rebuilding energy, and strengthening the immune system. In TCM, astragalus is used to reduce menopausal sweating and night sweats (Winston and Maimes, 2007, p. 148).

Immunity: Use in tea, as a powder, as a tincture, or in soup stocks throughout the fall and winter to strengthen immunity and prevent frequent colds, flu, bronchitis, pneumonia, and other externally contracted conditions. Use throughout the year if you live in a tick endemic area like Maine to strengthen immunity and lessen tick-bite symptoms. Keep the immune system strong and healthy with regular use of astragalus and supportive lifestyle practices. (pg 199 tips for gardeners during tick season).

Liver: Protects the liver against toxic substances, including carbon tetrachloride (Yance, 1999, p. 127) and prevents liver damage caused by pharmaceutical medications and viruses (Winston and Maimes, 2007, p. 149).

Lungs: Strengthens chronically weak lungs and treats shortness of breath. Astragalus is an important herb to rebuild the health and vitality of the respiratory (lung qi) and immune system following an acute infection such as bronchitis or pneumonia, and

Huang Qi Astragalus

Huang Qi Astragalus is in the Tonifying category in Chinese Herbal Medicine. Specifically, it tonifies the Spleen and Lung qi, tonifies the wei (defensive) qi, raises the Spleen and Stomach yang. The primary function of the Spleen is to transform and transport foods and fluids. The Spleen is the magician that transforms the whole of your life into something that nourishes you and then it transports that nourishment to all aspects of your being. So "food" could literally be a nice meal, and it is also a walk with a good friend, learning something new or appreciating an art form. It's all transformed into you by virtue of the Spleen. The primary function of the Lung is governing qi, sending it where it needs to go when it needs to be there. It includes the wei qi or protective qi (immune system) that moves just under the skin, discerning what is beneficial to let in and what is not. It's the receiver of the breath qi, or inspiration in all its forms. Huang Qi Astragalus is the Queen of Qi Tonics. Its movement is upward which enables it to bring qi all the way to your head and extremities engendering clear thinking and strength in the limbs. Its upright nature gives you the strength to digest all aspects of life, be informed by inspiration, and stand up and be who you are in the world. It's used to support digestive function and revive the energy; raise and hold things where they belong (i.e., prolapse or bleeding issues); strengthen the lung; stop inappropriate sweating and strengthen the immune system; facilitate urinary function and heal sores once they're past the really infected stage.

—AMY JENNER,
Acupuncturist and Chinese Herbalist

to prevent further infections. TCM states that the lungs help create an energy called *wei qi*. Chinese herbalists define *wei qi* as a kind of protective energy that prevents illnesses caused by exposure to external climatic influences such as cold, damp weather. Astragalus is an herb that supports *wei qi*.

Skin: Useful when taken orally in a soup, tincture, or decoction for people with first-, second-, or third-degree burns. It supports the skin's healing process and strengthens the immune system's response to trauma. I first saw how helpful astragalus is for serious burns after giving it to a family whose house burned down in the night. They fortunately all got out safely but not without the father being burned. Astragalus was one of the herbs we used to help his burns heal and to strengthen his lungs and immune system.

PREPARATION AND DOSAGE

Decoction: Simmer 4 tablespoons of the dried root in 16 ounces of water for 45-60 minutes. Steep 1–2 hours. Drink 2–3 cups per day. In the winter I often have a pot of water with roots such as astragalus simmering on my woodstove for 2–4 hours. My favorite wintertime tea, which I named Immune Tonic Tea (available from Avena Botanicals), contains the roots of astragalus, codonopsis, cooked rehmannia, cinnamon, ginger, and licorice. It tastes delicious and warms the body.

Soup stock: Simmer 8–12 tablespoons of freshly chopped or dried roots in 2 quarts of water for 2–4 hours. Freeze in ice cube trays. Store cubes in a glass container in the freezer. Use 1–2 cubes per cup of hot soup.

Powder: ½–1 teaspoon, mixed in hot water,

warm milk, hot cereal, or in a vegetable or fruit smoothie, 1–3 times per day.

Tincture: (1:4) Fresh or dried root, ¼–1 teaspoon, 3 times per day. We simmer the roots in water for 12 hours before adding the appropriate amount of alcohol. This two-step process extracts both the water-soluble and alcohol-soluble constituents from the roots.

Safety considerations: Not recommended during the acute stage of a cold, flu, fever, or pneumonia.

Long-term safety: Safe. Traditionally used as a medicinal food in China, cooked in soups and as a tea.

BURDOCK

Latin name: *Arctium lappa*

The genus Arctium is derived from the Greek *arktos*, "Bear." Many indigenous people revere the Bear as a great herbalist and collector of medicinal plants.

Common name: beggar's buttons

Family: Asteraceae

Place of origin: temperate Europe and Asia

Parts used: root and seed

Energy: cooling

Flavor: bitter, sweet, oily

Biennial: root is dug in the fall of the first year or early the following spring

Cultivating: If you wish to cultivate burdock (burdock can be cultivated, but it also grows freely and abundantly in fields and in a variety of places in the northern hemisphere where the soil has been disturbed), direct seed it 2–3 inches apart and ½ inch deep into raised beds in early spring as soon as the soil can be cultivated. Choose garden beds that are easy to dig and where the soil is deep. Some people use wooden sides when forming their raised beds, as burdock can grow a long taproot and be challenging to dig up. Seeds germinate in 1–2 weeks. Mulch young plants with straw.

Collecting: With a transplanting spade, dig the roots of the first-year plants in October or November, once the leaves have died back. Take good care to loosen the soil around the roots before attempting to unearth them. With extra patience the roots will emerge unbroken. Roots have a thin, light brown outer skin and are white and fleshy within. Wash the roots before storing or processing. They can be stored whole in a plastic bag[14] in a root cellar or refrigerator for several months and enjoyed in winter soups and teas. In Avena's apothecary we tincture burdock roots fresh. Roots can also be chopped and dried and used in wintertime and springtime teas and soups.

If you're digging the roots from nearby fields and meadows, be sure the environment is clean and safe (not an old dumping ground; check with the land owner or town) and that you have properly identified the plant before you engage in the digging process.

Actions: alterative, anti-inflammatory, diuretic, liver tonic, lymphatic tonic, nutritive

INDICATIONS

Digestion: Burdock's bitter taste stimulates digestive activity, clears toxins from the gut, lessens sweet cravings, and relieves gas, indigestion, and constipation. The root's mucilaginous qualities lubricate lower gastrointestinal tract tissue and help ease gastritis and irritable bowel syndrome. Burdock root contains inulin, a rich source of fructo-oliogosaccharides (FOS), which enhances the growth of healthy bowel flora. Burdock root has long been revered for its ability to remove accumulated waste from the tissues and bring the body back into a balanced state of health. Herbalist David Hoffman writes that burdock will "move the body to a state of integration and health" (2005, p. 528). Herbalist Matthew Wood writes, "Burdock helps to restore the primal blueprint of health, so to speak, when it has been lost in persons suffering from long, chronic illness" (2008, p. 105).

Gynecology: The sweet and oily properties of burdock root benefit the endocrine system and act as a nutritive tonic for women entering menopause and beyond. Useful in formulas for women with a prolapsed uterus.

Kidneys: The root's mild diuretic action on the kidneys helps clear the blood of harmful acids and reduce fluid retention, urinary calculi, and low back pain.

Liver: Improves overall liver function. The health of the liver is significant for women because it helps break down and eliminate excess levels of hormones like estrogen. A congested liver is less able to deal with peak levels of sex hormones, resulting in a wide range of discomforts such as breast tenderness, water retention, depression, frustration, anger, fatigue, and sluggish digestion. Burdock can be used consistently over a 1–3 month period to improve liver function, especially during the spring and fall months. It can be used a few times a week as ongoing support to the liver and digestive system.

---

14   Look for non-petroleum-based vegetable storage bags at a food coop or health food store.

Lymph: Burdock acts as a decongestant for lymph tissue; use as part of a formula for enlarged and tender lymph nodes, lymphedema, fibrocystic breast tissue, and mastitis.

Respiratory: Burdock can be combined with other herbs such as mullein leaf and flower and marshmallow root to reduce the dryness in conditions such as bronchitis and sinusitis.

Skin: A tincture of burdock seed can be taken internally for treating psoriasis and other dry, crusty, and scaly skin and scalp conditions. A burdock leaf poultice can be applied externally to heal wounds and ulcers.

PREPARATION AND DOSAGE

Decoction: Place 3–4 tablespoons of fresh or dried root into 3 cups of water, simmer for 15–20 minutes, then steep 30–40 minutes. Drink 1–3 cups per day.

Tincture: (1:4) Fresh or dried root. Take ¼–1 teaspoon, 3–4 times per day.

Soup stock: Simmer fresh or dried roots for 30–40 minutes in whatever amount of soup stock you are preparing, using 3–4 tablespoons of roots per 3–4 cups of water. The roots can be strained or eaten as part of a vegetable soup.

External use: A burdock root decoction can be used topically as a rinse for dandruff. A burdock leaf poultice can be applied externally to heal wounds and ulcers.

Safety considerations: Not recommended to use concurrently with insulin or with oral anti-diabetic agents because of possible increased hypoglycemia. Avoid during pregnancy and when breast feeding (McIntyre and Boudin, 2012, p. 20). Theoretically a person allergic to plants in the Asteraceae family may have a reaction to burdock leaf or root.

Long-term safety: Safe. Traditional use as a food.

Burdock Root and Fennel Seed Tea

Place 3–4 tablespoons of fresh or dried burdock roots into 3 cups of water and simmer for 15 minutes, covered. Add 1–2 tablespoons of fennel seed during the last 5 minutes of simmering. This combination is a favorite of mine to drink during the fall and winter months for warming the belly, supporting overall digestion, and easing gas and constipation.

DANDELION

Latin name: *Taraxacum officinalis*

Common names: lion's tooth, *pissenlis* (French), *diente de leon* (Spanish)

Family: Asteraceae

Place of origin: Asia and Europe. Today dandelion is commonly found throughout the temperate regions of the world.

Parts used: root, leaf, and flower

Energy: cool, moistening

Flavor: bitter, slightly sweet

Collecting: Dandelions are easily found in gardens, stone pathways, edges of farmers' fields, and meadows. This hardy, short-lived perennial offers extraordinary nutritional and medicinal benefit to humans and animals. The nectar of dandelion flowers feeds honeybees and the ruby-throated hummingbirds as they migrate north in spring. We dig dandelion roots in late April and early May, before they flower, and tincture the leaves and roots fresh. I also chop and dry some of the roots and eat lots of the fresh greens and flower buds in salads and vegetable dishes.

Actions: bitter, cholagogue, digestive and liver tonic, detoxifier, diuretic, laxative

Dandelion Pesto

¾ cup organic olive oil

3 cups fresh partially chopped dandelion greens

¼–½ cup organic walnuts, almonds, or sunflower seeds

2–4 garlic cloves

¼ tsp salt

Optional: ¼ cup of good-quality, freshly grated Romano cheese

You can mix dandelion greens with other greens such as fresh parsley, cilantro, chickweed, spinach, arugula, sage, and oregano. Enjoy foraging in the spring and fall garden for various greens. In the summer I like making pesto with garlic scapes.

First blend the olive oil, garlic, and greens until smooth and then add additional ingredients. Store in a glass jar with a non-BPA plastic lid in the refrigerator for up to 7 days. I like freezing pesto in half-pint glass canning jars with non-BPA lids. (Non-BPA plastic lids for narrow- and wide-mouth canning jars are available online and in some hardware stores that carry canning supplies.)

INDICATIONS

Bones: Dandelion leaves contain calcium, potassium, magnesium, manganese, iron, and boron and can be eaten fresh in salads and pesto, cooked in soups, and prepared as a vinegar extract with other mineral-rich herbs. Nutritive-rich herbs like dandelion can be used regularly in food and teas. Every spring and summer at Avena we make several gallons of a mineral-rich vinegar using locally produced organic apple cider vinegar as the base. We infuse fresh dandelion leaves, horsetail, chickweed, and nettle leaves in the vinegar along with an array of other fresh, mineral-rich herbs and call it Strong Woman, Strong Bones Vinegar.

Breasts: Use dandelion root as part of an herbal and dietary protocol for easing tender or congested breast tissue. The root's cooling and moistening qualities help to clear toxic heat from the body, making this herb beneficial for mastitis. Dandelion and echinacea root tinctures, ¼–½ tsp, taken every few hours, along with specific homeopathic remedies and cabbage leaf poultices, can be used for resolving mastitis.

Cardiovascular: Lowers LDL/VLDL cholesterol levels.

Digestion: Dandelion's bitter taste stimulates saliva and HCL (hydrochloric acid) production, bile flow, and digestive enzymes. This action improves poor fat metabolism and sluggish digestion, increases the flow of digestive juices, enhances assimilation and elimination, cleanses the liver, and eases indigestion and gas. The root contains inulin, a source of fructo-oligosaccharides (FOS), a probiotic that stimulates the growth of healthy intestinal flora. Herbalist Juliette de Bairacli Levy recommends eating the mineral-rich greens for strengthening the enamel of teeth and for improving the health of people with diabetes, obesity, and over-sleepiness (Levy, 1973, p. 57).

Gynecology: I was taught to add dandelion root into an herbal protocol for women who have various menstrual and menopausal symptoms related to hormonal shifts and imbalances, especially conditions associated with excess estrogens such as endometriosis. Liver-supporting herbs like dandelion are beneficial for women coming off birth control pills and for easing premenstrual stress and peri-menopausal and menopausal symptoms. As a gentle laxative, dandelion root moistens the stool, which can be helpful for any woman experiencing constipation. Dandelion root combined with fennel seed, milk thistle seed, and blessed thistle increases milk production.

Kidneys: Dandelion leaf contains potassium and is an effective diuretic that is nonirritating to the kidneys and urinary tract. (Most pharmaceutical diuretics contribute to a loss of potassium.)

Liver: Dandelion is my favorite herb for improving overall liver function. As a general liver tonic it can be taken daily for a 1–3-month period each spring and fall. It's an important herb for reducing swelling and congestion of the liver and gallbladder. Dandelion root can help resolve jaundice. It's useful as an overall tonic following surgery or a chronic illness (use with other tonic herbs specific to the individual). David Winston recommends combining dandelion root with milk thistle seed, schisandra berries, and turmeric root for people with hepatitis. Dandelion root along with other hepatic herbs such as yellow dock root, milk thistle seed, artichoke leaf, motherwort, and turmeric improves the phase I and phase II detoxification pathways (Romm, 2010, p. 46).

Skin: Dandelion root combined with burdock root and burdock seed helps resolve acne, boils, and abscesses.

PREPARATION AND DOSAGE

Decoction: Simmer 1–2 tablespoons of fresh or dried roots, chopped, in 8 ounces of water for 20–30 minutes. Steep another 30–45 minutes before drinking. Drink 1–2 cups per day.

Infusion: Infuse 2–3 tablespoons of freshly chopped leaves in 8 ounces of hot water for 5–10 minutes. Drink 1–3 cups per day.

Glycerite: (1:5) Fresh or dried root. Take ¼–½ teaspoon, 2–3 times per day.

Tincture: (1:5) Fresh root, fresh root and leaf, dried root. Take ½ teaspoon, 2–3 times per day.

Oil: Collect newly opened flowers in the morning once they're no longer damp with dew. Fill a clean glass pint-sized or quart jar with flowers, gently pressing them into the jar. Then fill the jar completely with an organic vegetable oil such as olive, almond, or apricot kernel, cover with unbleached cheesecloth, and place in a warm sunny window for 7–10 days. Strain the flowers from the oil and store in a glass jar in a cool, dark cupboard for up to one year. Makes a lovely massage oil.

Ode to the Dandelion

There is nothing more delightful for me in spring than a hillside ablaze with yellow dandelion flowers. Every spring I marvel at these flowers in my garden—each flower a perfect mandala, alive with buzzing insects of all kinds. Dandelion flowers continue to teach me to take nothing for granted. Though I have collected hundreds of pounds of dandelions in my lifetime, they continuously remind me not to tire of the constant and unique beauty each plant and each moment offer. And in the same breath as I say thank you to the dandelion, I whisper a prayer for the well-being of the people who spray herbicides on their lawns to eradicate dandelions and other weeds.

May all beings be safe and protected. May all beings be at peace.

In 1990, over 30 million pounds of insecticides were applied to residential gardens and lawns in the United States (*Wall Street Journal*, October 30, 1991). In 2012, insecticides, pesticides, and herbicides continue to pose serious health threats to animals, insects, earthworms, humans, and numerous ecological

Safety considerations: Use cautiously with antihypertensive and diuretic drugs. Theoretically a person allergic to plants in the Asteraceae family may have a reaction to dandelion flower, leaf, or root.

Long-term safety: safe

SOLOMON'S SEAL

Latin names: *Polygonatum biflorum, P. biflorum var. commutatum, P. multiflorum* (native to Europe)

Common names: true Solomon's seal, giant Solomon's seal

Family: Ruscaceae, formerly in the Liliaceae Family

Place of origin: North America (eastern deciduous forests of U.S. and Canada), other species are found in Asia and Europe

Parts used: rhizome

Energy: cool, moist

Flavor: sweet, slightly acrid

Cultivating: Fresh (undried) seeds can be sown into a prepared seed bed (forest soil) in the fall. Seeds need to freeze and thaw through the winter and early spring months and can

tems. The U.S. Environmental Protection Agency estimates that today about 80 million U.S. households dump rly 90 million pounds of herbicides and pesticides on lawns in a year.

With lawns covering almost 50 million acres in the United States, the impact this has on our health and environment tally unacceptable and yet this and other horrendous acts against the natural world occur daily. U.S. lawns are sed with about 270 billion gallons of water a week — enough to water 81 million acres of vegetables — all summer g. More chemicals per acre are used on lawns than in agriculture. With 100 million pounds of pesticides used by neowners each year, that's up to 10 times as much pesticides as are used on farmland. Add in the 90 million pounds erbicides applied on lawns every year, and you've got a deadly brew.

Breyten Breytenbach, South African poet and author of *True Confessions of an Albino Terrorist,* once said to writer y Tempest Williams:

"You Americans, you have mastered the art of living with the unacceptable."

(from "The Politics of Place: An Interview with Terry Tempest Williams" by Scott London)

take a few years to germinate. Horizon Herbs sells fresh, pre-stratified seeds. This perennial plant is hardy to -40°F. I have found it simpler and easier to purchase seedlings from a reputable native plant nursery. My first seedlings came from Prairie Moon Nursery in Winona, Wisconsin. True Solomon's seal is commonly found growing around old New England farmhouses. I have tried transplanting older roots without much success. The Solomon's seal bed growing in Avena's garden receives morning shade and afternoon sun and is protected from the deer, who enjoy munching the plants' luscious spring stalks. The rhizomes have spread thickly over several years, creating a mat-like bed. Richo Cech from Horizon Herbs says, "Root restriction is relished."

I am encouraging lots of organic and biodynamic herb growers to cultivate beds of Solomon's seal. Herbalists and herbal medicine makers desperately need more organic sources of this root as it has so much medicinal value!

Collecting: I waited seven years before collecting a handful of rhizomes from Avena's garden. The rhizomes are harvested in the fall once the stalks have dried up and the berries have dropped. I prefer to tincture the rhizomes fresh. For making oil we chop the rhizomes and dry wilt them in a basket for 24 hours. We then add the rhizomes to organic olive oil and infuse them at 100°F for 2 weeks in an enamel oven roaster.

Herbalist Karyn Sanders recently told me that she was instructed by her native elders to only use a wooden digging stick when collecting Solomon's seal root and to be extremely careful when untangling the rhizomes from each other. She emphasized how important it is to not break any of the rhizomes during the harvesting process and "that you can't leave even a hair in the ground or it will break it's energy".

Matthew Wood writes about Solomon's seal in *The Book of Herbal Wisdom,* "The horizontal rhizome under the ground, with the vertical stalk rising up from it indicates a wolf medicine. . . . The Wolf Medicines usually have a ninety-degree angle in their construction, indicating an affinity to making profound changes or turns in life. They help bring a person to a transformative place, or help them go through a change, or help them adapt to a change that has already occurred. The ninety-degree angle represents joints in the organism, and key-joints in the path of life (Wood, 1997, pp. 402, 404).

Actions: qi tonic, demulcent, kidney tonic, nutritive, rejuvenative, restorative to the muscular system

INDICATIONS

Digestion: Nutritive and lubricating. The rhizomes moisten and soothe the mucosa lining of the intestines and relieve dry constipation and gastritis. A restorative tonic for people whose digestive tracts are weak due to poor-quality food and chronic stress.

Gynecology: I recommend women experiencing menopausal changes such as mood swings, a loss of identity or direction, hot flashes, or anger and agitation to use this herb regularly over several months. Solomon's seal is cooling, moistening, and healing to dry vaginal tissue and cervicitis when taken orally as a tea or tincture or as a powder mixed in warm water. Combine Solomon's seal with other herbs (astragalus and burdock root) to prevent or support a woman with a prolapsed uterus.

Musculoskeletal: Reduces inflammation, tightness, and pain. Strengthens, tones, lubricates, and repairs injuries to the tendons, ligaments, attachments, and joints. I have found taking the root tincture over several months to resolve my own and other gardeners' tendonitis (located in the forearms), to resolve overly tight or overly loose tendons, ligaments, attachments, and joints, and to relieve repetitive-use injuries such as carpal tunnel syndrome. I also add Solomon's seal root into a formula for broken bones. Herbalist David Winston says it helps nourish cartilaginous tissue and heal old, poorly healed trauma injuries. It can be of benefit for torn rotator cuffs, compressed discs, mild tears of the meniscus, and ACL and sacroiliac joint pain.

Respiratory: The rhizome's lubricating and demulcent properties soothe dry, irritated lung tissue. Combine astragalus and burdock root with Solomon's seal for a person experiencing a chronic cough and a depleted immune system. Lingering coughs and fatigue often follow an acute flu or respiratory infection. Use the roots for at least one month, or longer, to restore health and vitality to the lungs and immune system.

PREPARATION AND DOSAGE

Decoction: Soak 4–6 tablespoons of chopped rhizomes overnight in 16 ounces of cool water. Then simmer for 20–30 minutes. Steep another 30–45 minutes before drinking. Drink 1–2 cups per day.

Powder: The mucilaginous roots are not easy to powder. A special grinder is needed so as to not gum up the rotating blades. Mix ½–1 teaspoon of the powder into warm milk or water, or cooked cereal. I like adding the powder to my morning blueberry smoothie. I recommend that people who use Solomon's seal for longer than 2–3 months take the powder instead of the tincture.

Tincture: (1:5) Fresh or dried roots. Take ¼–½ teaspoon, 2–3 times per day. Herbalist Matthew Wood recommends smaller doses, 1–5 drops daily.

Oil: Freshly dried roots can be infused in an organic vegetable carrier oil for two weeks at 100°F. Strain, bottle, and label the oil. Use as often as needed for musculoskeletal conditions such as tendonitis, bursitis, arthritis, frozen shoulder, and sprains and strains.

Safety considerations: Berries are toxic.

Long-term safety: safe

BELOW: Photo of an outdoor painting taken by the author in a park in Sicily.

BLESSED THISTLE

Latin names: *Cnicus benedictus* (Gaetn.), *Carduus benedictus* (Steud.), *Carbenia benedicta* (BERUL.)

Common names: holy thistle, St. Benedict thistle, Cardo Santo thistle

Family: Asteraceae

Place of origin: southern Europe

Parts used: leaf and flower

Energy: cooling

Flavor: bitter, pungent

Cultivating: This annual herb is easily propagated by seed. The seeds are good sized, ¼ inch, and easy to handle. Sow the seeds in spring, one month before last frost date, into seedling trays. Cover with ¼–½ inch soil. Germinates quickly—within 4–5 days. Transplant into individual 3–4-inch pots once the seedlings are 2 weeks old. Plant outside in a well-prepared garden bed after the last frost date. Seedlings prefer to be planted 2 feet apart in ordinary and well-drained garden soil (not too rich). Water daily during the first week after transplanting. Seeds can be collected in the fall, cleaned, and stored in an envelope or glass jar in a dark, cool, dry cupboard.

Collecting: Take extra care when collecting the leaves and flowers for this herb is quite prickly. The Felco #310 harvest shears work well when cutting the flowering stalks of blessed thistle. We begin collecting this herb when it is bushy and flowering, early to mid-July, and continue harvesting through late August. We tincture the leaves and flowers fresh and lay some out to dry for tea. We also let many flowers go to seed so that later in fall there are plenty of seeds to save for next year's garden.

Actions: bitter tonic, cholagogue, circulatory stimulant, diaphoretic, emetic, emmenagogue, galactagogue

INDICATIONS

Digestion: Small doses of this bitter-tasting herb taken before each meal helps to increase secretions of saliva, gastric juices, pancreatic enzymes, and bile. A few drops of the tincture under the tongue stimulates appetite, improves energy and vitality, and eases indigestion, bloating, nausea, and gas. Blessed thistle can be used alone or in combination with other bitter-tasting herbs. Avena Botanicals Bitters Tonic contains blessed thistle, artichoke leaf, burdock root, licorice root, and ginger root.

Gynecology: Helps regulate hormones in girls at puberty (Lad and Frawley, 1986, p. 62). Its cooling properties soothe anger, irritability, and eruptive emotions before and during menstruation and menopause. Can be combined with other herbs to ease cramps and painful menstruation. Helps reduce menstrual and menopausal headaches. Blessed thistle promotes lactation. For nursing women I prefer to combine blessed thistle with fennel seed, nettle leaf, red raspberry leaf, and peppermint in tea.

Head: Increases circulation to the brain. Traditionally used to improve memory and poor concentration. Eases headaches caused by a sluggish and congested liver.

Liver: Useful for people with hepatitis and jaundice. I often combine blessed thistle with milk thistle seed and fennel seed for improving liver health and supporting a smooth flow of qi in the liver. These three herbs offer support to women who suffer from headaches, hormonal swings, indigestion, gas, and liver congestion—especially following birth and during menopause.

PREPARATION AND DOSAGE

Infusion: 1–2 teaspoon dried herb in 8 ounces hot water, steep for 5–10 minutes. Take ¼–½ cup, 15–30 minutes before meals.

Tincture: (1:4) Fresh or dried leaves and flowers, ¼–½ teaspoon, 2–3 times a day, before meals.

Safety considerations: Avoid during pregnancy. Allergic reactions may occur for people sensitive to plants in the Asteraceae family. Ingesting large amounts of blessed thistle may cause vomiting.

Long-term safety: safe

Blessed Thistle as a Heal-All

Maude Grieve writes in *A Modern Herbal* that blessed thistle received its name from being highly valued as a *heal-all*, including curing the plague. "It's mentioned in all the treatises on the Plague, and especially by Thomas Brasbridge, who in 1578 published his Poore Man's Jewell, that is to say, a Treatise of the Pestilence, unto which it annexed a declaration of the virtues of the Hearbes Carduus Benedictus and Angelica" (Grieve, 1931, p. 795). Blessed thistle was commonly cultivated in European monastic gardens.

GOTU KOLA

Latin name: *Centella asiatica*

Common names: Indian pennywort, Brahmi, Mandukaparni

Family: Apiaceae

Place of origin: India, Sri Lanka, and Southeast Asia

Part used: leaf, aerial parts

Energy: cooling

Taste: bitter, slightly sweet

Cultivating: This special perennial herb can be started from seed by adventurous gardeners who have a greenhouse, heat mat, or warm windowsill. Gotu kola seeds need 70–75° temperatures to germinate. I wait until the second week of June before transplanting seedlings into the warmest and sunniest location in the garden. Gotu kola loves to grow in a sunny garden bed near a stone wall where the stones warm up and the plant can send its runners over the wall. Seedlings can be purchased from Horizon Herbs and Richter's.

In September, before our first frost, I cut back and dig up several gotu kola plants and place them into large shallow pots. These pots winter indoors in sunny windows. They need to be watered 2–3 times a week. Be aware of aphids during the winter months when the plants are tired of being inside. In April I divide the plants and put them into bigger pots in preparation for the coming growing season. Gotu kola grows as a perennial in the tropics and as an annual in northern climates.

Collecting: Throughout the summer I nibble on 3–5 leaves each day. Ayurvedic practitioners say eating a few fresh leaves every day enhances memory. For drying or tincturing, we harvest the sprawling leaves and flowers, 2–3 times during the summer months. We tincture the plant fresh and infuse fresh leaves in organic sesame oil. The fresh leaves infused in oil create a very special massage oil, highly regarded by Ayurvedic practitioners.

Actions: alterative, anti-inflammatory, antipyretic, anxiolytic, memory tonic, nervine, rejuvenative, vulnerary

INDICATIONS

Circulation: Gotu kola stimulates microcirculation and healing to tissues that have been traumatized, including post-surgical trauma. It improves the integrity of capillaries, making it useful for the eyes, venous insufficiency, spider veins, varicose veins, and anal fissures (McIntyre and Boudin, 2012, p. 204). I recommend combining fresh gotu kola tincture with hawthorn berry tincture for strengthening capillaries and veins.

Immunity: In his book *Herbal Medicine, Healing and Cancer,* Donald Yance writes that gotu kola "increases the formation of connective tissue components such as hyaluronan (also called hyaluronic acid), a natural substance produced by cells. When large amounts of hyaluronan are produced, it can block the signals of ras, a cancer-causing gene, and stop the growth of tumor cells" (Yance, 1999, p. 100).

Gotu kola clears toxins and inflammation and has traditionally been used for easing arthritis, gout, and joint inflammation. Gotu kola can be part of a protocol for people with Lyme disease when their symptoms include joint inflammation and memory loss.

Mind: Gotu kola enhances concentration and alertness, improves memory, and relaxes the brain when over-stimulated (Yance, 1999, p. 99). Anne McIntyre recommends gotu kola for calming the mind and assisting people whose minds are chaotic. In the Ayurvedic tradition gotu kola is used to strengthen the mind, slow the aging process, support people with Alzheimer's disease, and assist people who have experienced a stroke. Gotu kola can be used by people who meditate to help quiet the mind and cultivate a harmonious and compassionate state of being.

Nerves: Calms and relaxes the central nervous system. Reduces anxiety, depression, and insomnia. Useful for people with behavioral and developmental difficulties such as autism, Asperger's, and ADHD (Pole, 2006, p. 187). I combine gotu kola with lemon balm and hawthorn berry for reducing anxiety and relaxing the nervous system.

Skin: Enhances wound healing and reduces formation of scars. Beneficial for inflammatory conditions such as eczema, psoriasis, acne, urticaria (hives), and burns. Donald Yance recommends using gotu kola both orally and topically in postoperative cancer surgery and post-radiation therapy due to its ability to stimulate the rapid growth of the reticuloendothelial system, increasing healing and decreasing inflammation. Can be taken orally for connective tissue disorders such as scleroderma, cellulite, varicose veins, hemorrhoids, and lupus (Kuhn and Winston, 2001, p. 164). Sebastian Pole, author of *Ayurvedic Medicine,* writes: "It increases the rate of keratinisation of the skin, which helps to strengthen the outer protective layer of the body" (Pole, 2006, p. 187).

PREPARATION AND DOSAGE

Food: Eat a few fresh leaves daily. Botanist Nishi Rajakaruna from Sri Lanka described to me how he enjoys eating red rice cooked in coconut milk with freshly chopped gotu kola leaves as a breakfast dish.

I make a summertime smoothie with a handful of fresh gotu kola leaves, organic blueberries, almonds, and red lycii berries.

Mix ½–1 teaspoon high-quality organic gotu kola powder into a small amount of ghee and eat with steamed greens—an excellent way to rejuvenate and restore cerebral function (Pole, 2006, p. 188).

Infusion: Infuse fresh or dried leaves in cool water overnight and drink at room temperature the next morning. Heat can destroy gotu kola's properties. Dried leaves lose their medicinal properties within one year of being harvested (Pole, 2006, p. 188).

Powder: Use high-quality, organic powder that has recently been dried and powdered. Can be mixed in a smoothie with coconut milk, almond milk, or organic raw milk.

Tincture: (1:2) Fresh leaves, ¼–½ teaspoon, 3 times per day.

Topical: For washing wounds, use as a cool infusion or dilute tincture in cool water. As a fresh poultice, chop fresh leaves and place over hot, inflamed area and secure with a cotton bandage. Anne McIntyre recommends mixing gotu kola with coconut oil for relieving skin conditions such as eczema, herpes, acne, and boils (McIntyre and Boudin, 2012, p. 204).

Safety considerations: Avoid during pregnancy. Avoid if taking barbiturates or benzodiazepines, as gotu kola can inhibit liver enzymes responsible for barbiturate metabolism (Poole, 2006, p. 188). Avoid if using insulin/hypoglycemic and cholesterol lowering medications. Use cautiously when using central nervous system (CNS) depressants: alcohol, opiates, anaesthetics, tricyclic antidepressants, and anti-epileptics (McIntyre and Boudin, 2012, p. 204).

Long-term safety: safe

Herbology and nutrition are a single science in Ayurveda, and no treatment can be truly efficacious that neglects one or the other. Food deals with the grosser nutrition of the body; herbs give subtle nutrition and stimulation to the deeper tissues and organs. (Lad and Frawley, 1986, p. 35)

Latin name: *Melissa officinalis*

Common name: lemon balm

Family: Lamiaceae

Place of origin: Mediterranean

Part used: leaf and flower

Energy: cool, moist

Flavor: sour, mildly bitter

Cultivating: This perennial herb is easy to propagate from seed. Seeds need light to germinate, so do not cover with soil when planting. We press 3–4 seeds into individual plugs and gently water them regularly so as not to dislodge the seeds before they germinate. Once the young seedlings reach 2 inches in height we place them into 4-inch pots, where they can develop larger and stronger roots. I have observed over several years that lemon balm flourishes in partial sun as long as the plant receives a fresh layer of compost every spring. We transplant our seedlings in late May or early June into composted garden beds that are mulched with straw. We keep the seedlings well watered while they're establishing themselves. I rarely cover the plants with straw in winter but have heard that gardeners in zones colder than zone 5 need to cover their plants.

Collecting: By mid-June we harvest the leaves from our well-established perennial plants and immediately tincture them in organic alcohol and organic vegetable glycerin. We lay hundreds of pounds of leaves to dry on screens in our drying room. We usually get a second harvest of leaves in mid-August if enough rainfall has occurred during the summer. Erratic weather patterns require me to pay more attention to plants that may need extra water or mulch.

Actions: antiviral, antidepressant, antispasmodic, carminative, diaphoretic, memory aid, nervine, thyroxine inhibitor

INDICATIONS

Digestion: Calms a nervous stomach and stress-related digestive upsets, including heartburn. Eases nausea, gas, bloating, and vomiting.

Endocrine: Lemon balm is beneficial for people with an overactive thyroid. The leaves act as a thyroxine antagonist. Holistic health care providers use lemon balm with motherwort and bugleweed for people with hyperthyroidism and Graves' disease.

Gynecology: Lemon balm helps ease painful menstruation and lessens the severity of hot flashes. I have used a tincture (called Into the Flow) combination of fresh lemon balm, lavender, blue vervain, rose petals, and milky oat seed over the last 20 years for women challenged by premenstrual and menopausal agitation and stress. I am forever grateful for the comfort and relief that herbs offer women.

Heart: Eases heart palpitations associated with stress. Combine with hawthorn flowers, leaves, and berries and motherwort for lowering mildly elevated blood pressure. Lemon balm calms and quiets the heart.

Immunity: As an antiviral, use orally and topically to prevent and treat outbreak of herpes, chicken pox, shingles, fevers, and colds.

Mind: Lemon balm has long been revered by European herbalists for its ability to enhance memory and circulation to the brain. Research has shown that lemon balm interferes with cholinesterase, which breaks down acetylcholine. Using lemon balm to stimulate this process may help in lowering the incidence or slowing the progression of Alzheimer's disease (Perry N. et al., 1996, *International Journal of Geriatric Psychiatry:* 11(12): 1063-1069). I encourage people who have a family history of memory loss to consider using lemon balm several times a week or in combination with herbs such as bacopa (an Ayruvedic herb that can be grown as an annual in northern climates), gotu kola, rosemary, and sacred basil.

Nerves: Eases stress, anxiety, panic attacks, insomnia, attention deficit disorder (ADD) and attention deficit hyperactivity disorder (ADHD), and general nervousness. I especially like to use this herb for children and adults who easily become overwhelmed, who are considered highly sensitive, or who have low self-esteem. Lemon balm relieves tension headaches, premenstrual and menopausal stress, irritability, and depression. Lemon balm combined with motherwort and rose petals is nourishing and supportive for a new mother. Lemon balm is safe and effective for reducing restlessness and hyperexcitability in children with attention deficit disorder (Kuhn and Winston, 2001, p. 211). I like to combine lemon balm with lavender, St. John's wort, rose petals, and calendula flowers, in a tea or tincture, for uplifting a person's spirit, especially in winter.

PREPARATION AND DOSAGE

Aromatherapy: An important essential oil for having in your home medicine chest and travel bag. Placing a drop on your pillow at night promotes a more restful sleep. Use a few drops in a foot bath or full body bath to relax the body and quiet an overwhelmed mind. I recommend that women working in stressful environments keep a small bottle of the essential oil in their bag or purse. Smelling the oil throughout the day helps calm the mind and connects a person with the healing forces of the plant world.

Bath: During the summer I place handfuls of fresh lemon balm along with various flowers into my bathtub and enjoy soaking amidst the floating herbs. A few drops of the essential oil can also be placed into the bathtub or shower for deep relaxation.

Glycerite: (1:3) Fresh leaves ground in organic vegetable glycerin. Use a few drops up to ½ teaspoon, 1–5 times per day.

Tincture: (1:3) We harvest the fresh, lush leaves in early summer and tincture immediately in organic alcohol. Use a few drops up to ½ teaspoon, 1–5 times per day.

Infusion: To make a warm infusion, place a handful of fresh leaves in a pot, cover with water, and slowly warm—do not bring to a boil. Keep covered and steep for 10–20 minutes. I often fill a quart glass jar with fresh lemon balm leaves, cover with cool water, and sip throughout the day during the summer months. Lemon balm is cooling to the body. Carefully dried leaves have much less of a lemony flavor but can still be made into a hot infusion. Place 3 teaspoons into a cup of hot water. Steep covered for 20 minutes. Drink 2–3 cups per day. Lemon balm can be combined with other fresh or dried herbs such as calendula, lady's mantle, nettle, oat seeds, St. John's wort, and sacred basil.

Infused medicinal oil: We harvest the fresh, lush leaves once the dew has dried and lay them out in baskets to dry wilt for 6 hours. The process of dry wilting reduces the moisture content. We then place the leaves into organic olive oil and infuse them at 100°F for two weeks. This oil can be used directly on cold sores, herpes, shingles, and hives and made into a salve or cream for these conditions.

Safety considerations: Avoid using lemon balm if taking thyroid medications for hypothyroidism. Considered to be a thyroxine antagonist. If pregnant, consult your midwife before using.

Long-term safety: safe

Lemon Balm Cordial

My garden mentor and friend Adele Dawson loved making herbal cordials. Lemon balm cordial was one of her favorites. The following recipe is a variation on Adele's recipe.

Chop several handfuls of fresh lemon balm leaves and fill a glass jar. Cover the leaves with a fifth of organic vodka or high-quality brandy, three-quarters of a cup of raw honey, and grated organic lemon peel. If you want an even stronger lemon balm flavor, place all the ingredients in a blender and add more fresh lemon balm leaves. Pour into a glass bottle with a lid. Shake well once or twice a day for one week. Strain into a clean glass bottle with a lid. Herbal cordials make lovely holiday gifts, given in smaller, 4-ounce glass bottles.

NETTLE

Latin name: *Urticaria dioica*

Common name: stinging nettle

Family: Urticaceae

Place of origin: Native to Europe. Naturalized in temperate climates worldwide.

Part used: young leaves, root, seed

Energy: cool, dry, neutral

Flavor: salty, young leaves are mildly sweet

Cultivating: This hardy perennial can be grown from good-quality seed, though I have found it much easier to establish a nettle bed from root cuttings. My first nettle plants arrived by chance in a pile of fresh cow manure delivered by a local farmer in his truck. One of the best gifts an herbalist could ask for! Nettle loves growing in moist, fertile soil and in old compost piles. As long as nettle has enough moisture and fertility, it will flourish in the sun or partial shade. Every few years, later in the fall, we shovel compost on top of Avena's nettle beds. Tree bark chips placed between our rows of nettles maintain good walking paths for us gardeners. This practice keeps the nettles from spreading into our walkways.

Collecting: Nettles are best collected in the spring when they reach about 10–12 inches high. We aim to have our harvest finished by the third week of May, before the red admiral butterfly arrives and lays its eggs on the plants. The hungry black caterpillars of the red admiral butterfly will devour a nettle patch within a few days. If this happens to you, take a scythe and cut the stalks down to the ground. Nettles will grow back. Our nettle beds come back thick and green in September. I drink lots of fresh nettle tea and cook with the leaves until a really hard October frost burns the leaves, announcing the close of the growing season.

When collecting nettles, our outfits include long-sleeved cotton shirts, pants, and gloves made for pruning thorny rose bushes. We use an Asian cutting tool called a kama, which works well for the several hundred pounds of nettles we harvest. We lay large cotton sheets onto the ground and directly cut the plants onto them. Once the sheets are full, we carry them into Avena's fresh plant receiving area, where we weigh what is needed for our fresh tinctures and take the rest to our drying room.

Actions of leaf: adaptogen, alterative, antihistamine, anti-inflammatory, astringent (mild), capillary tonic, diuretic, galactagogue, nutritive tonic

Matthew Wood writes: "It's one of the plants highest in protein and helps all protein pathways in the body—digestion, immune response, liver metabolism, skin reactions, and kidney elimination. It contains chlorophyll, indoles (including histamine and serotonin), acetylcholine, flavonoids, vitamins (including C), proteins, and dietary fiber" (Wood, 2008, p. 497).

Actions of seeds: Improves thyroid function and reduces goiters.

Actions of roots: Has an anti-inflammatory and decongestant action on the prostate gland.

INDICATIONS

Blood: The leaf is high in minerals and micronutrients, especially iron, calcium, magnesium, potassium, and protein. Taken regularly, nettles restores the body's overall energy and vitality and is effective for anemia, strengthening capillaries, preventing nose bleeds, and preventing skin that is thin and papery from bruising easily.

Endocrine: Nettle seed helps improve low thyroid function (hypothyroidism) and reduce goiter (McIntyre, 1994, p. 39). The leaf and seed taken together as tea or tincture are beneficial for people with adrenal fatigue. I encourage people who are chronically tired to use nettle leaf and seed together in a daily tea and to combine nettle seed with sesame seeds and salt as a food condiment.

Endocrine system, male: David Winston recommends combining nettle root with saw palmetto, white sage, and collinsonia root for benign prostatic hyperplasia (BPH).

Gynecology: Nettle leaf is safe and nourishing for pregnant and nursing women. It stimulates milk production. Nettle leaf is supportive to women whose hemoglobin/hematocrit levels are low. Its astringent action and nutritive qualities are helpful for women who experience a heavy menstrual flow or who feel tired, lethargic, and stressed. Nettle leaf as tea and cooked in soups is a nutritive tonic and rejuvenative remedy for menstruating and menopausal women, nursing mothers, and women beyond menopause.

Immunity: Regular use of the leaf tea, fresh tincture, or freeze-dried capsules relieves seasonal allergies, especially when started a month before the allergy season begins.

Kidneys and bladder: Nettle leaf and seed are an excellent tonic to the kidneys. The leaf has long been known to have a stimulating effect on the kidneys and bladder, supporting the body's elimination of metabolic wastes and uric acid. It's an herb to consider using to prevent urinary stones and calculi. Nettle seeds in particular are known to help restore health and strength to the kidneys.

Musculoskeletal: Nettle leaf is beneficial for people challenged with gout, arthritis, and osteoarthritis. Matthew Wood recommends nettle leaf for atrophy and paralysis of muscles (voluntary and involuntary). It's a valuable herb to give to people whose muscles are weak following surgery, the removal of a cast, or a prolonged illness.

Nutritive tonic: Herbalists consider nettle to be an invaluable nutritive tonic because it is high in vitamins, minerals, and protein. Juliette de Bairacli Levy used nettle "to cleanse the blood, tone up the whole system; as a cure for anaemia, rheumatism, sciatica, arthritis, obesity, infertility" (de Bariacli Levy, 1974, p. 103). Many herbalists recommend using nettle for a person who is run down following an illness, surgery, or from daily stress. I drink a quart of fresh nettle tea every day from late April through May along with adding it into various spring soups. Fresh nettle renews my overall energy and gives me the strength to garden 10–12 hours/day. In winter when I long for a taste of the summer garden I make tea from dried nettle with calendula blossoms, lady's mantle, oat seed, rose petals, and sacred basil.

Skin and hair: Juliette de Bairacli Levy taught me to use nettle tea as a rinse on dogs who have red, irritated hot spots. Nettle leaf tea or nettle leaf vinegar can be used as a hair rinse for humans to improve hair color and texture and remove dandruff. Nettle taken orally—tea, tincture, freeze-dried capsules—is useful in chronic skin problems such as eczema, boils, abscesses, and hives, and it also addresses food sensitivities and other digestive-related issues. Nettle tea can be used topically as a rinse for healing burns, sunburns, and insect stings.

PREPARATION AND DOSAGE

Infusion: Collect the fresh young leaves in the spring (and fall), place in a non-aluminum pot, cover with water, and bring to a simmer. Some herbalists simmer the tea for 15–20 minutes, while others bring the tea to a simmer, turn off the heat, and let it infuse for 30–60 minutes. If preparing with dried leaves, use 1–3 tablespoons of dried leaves per cup of water. Steep in hot water for 30–60 minutes. Drink 2–4 cups per day. Nettle can be combined with a variety of herbs such as lady's mantle, lemon balm, red raspberry leaf, red clover, mint, calendula, rose petals, sacred basil, and alfalfa.

Fresh leaf tincture: (1:3) ½–1 teaspoon, 3 times per day

Fresh root tincture: (1:4) ½ teaspoon, 3–4 times per day

Seed tincture: (1:4) ½ teaspoon, 3–4 times per day

Vinegar: Infuse fresh leaves in organic apple cider vinegar for one month. Strain and store in a glass jar. Some herbalists dilute the vinegar in water before using as a hair rinse, and others use the vinegar undiluted. Nettle vinegar is great on salads and steamed greens.

Safety considerations: Some people develop a red, irritated rash from skin contact with the plant. Crush the plants of fresh jewelweed, plantain leaves, yellow dock leaves, or chickweed and place onto a nettle rash to alleviate the irritation. Avoid using nettle for people with edema that is caused by impaired cardiac or renal function.

People who tend to be cold and have dry skin, hair, and eyes should use only small doses of nettle so as not to aggravate these conditions. Consult your health care provider if taking insulin, anticoagulants, antihypertensives, or diuretic medication, as nettle could possible potentiate their action. Nettle is not recommended for people with hemochromatosis.

Long-term safety: safe

Nettle Soup with Lovage and Rosemary

This is one of my favorite spring soups. Nettle and lovage are the bringers of spring in my garden. I love making nettle soup, and each time it's always a bit different.

Collect a basket of young, fresh leaves. Cooked nettles shrink a lot, so don't be shy about picking lots of large handfuls. Cut 2–3 stalks of lovage if you have this spring perennial in your garden.

First make a soup broth by placing 2–4 Tbl of dried or fresh rosemary into a pot with 8–12 cups of water and cover with a lid. Bring to simmer. Turn off the heat and let the soup stock infuse for 30 minutes.

Saute 2–3 organic onions (and leeks if you have them) in ghee or organic olive oil. Chop and add the lovage when the onions are soft, and cook together for a few minutes. Add as much fresh garlic as you like.

Strain the rosemary from the soup stock and add the fresh nettle leaves, using a wooden spoon to stir and mix the leaves with the soup stock. Add the onions and lovage and simmer on low for 5–10 minutes. Use a blender to create a thick green soup. Season with salt and freshly ground pepper and enjoy!

Variations: Sometimes I add potatoes and carrots to create a thicker soup.

Love by the way you walk, the way you sit, the
way you eat. This world very much needs love.
—THICH NHAT HANH

ROSEMARY

Latin name: *Rosmarinus officinalis*

Common name: dew of the sea

Family name: Lamiaceae

Place of origin: Mediterranean

Part used: leaf and flower

Energy: warming, dry

Flavor: pungent

Cultivating: I love rosemary. For over 27 years I have carefully cared for several large rosemary plants. This is a challenge in Maine, as we have to dig them out of the ground in late October, place them in large plastic pots, and wheelbarrow them indoors where they winter until mid-April. Rosemary plants can survive mild frosts. They just cannot live outside where the ground freezes solid in winter. Once indoors, I water them well once or twice a week. The Latin name of rosemary, *Rosmarinus,* means "dew of the sea." Rosemary loves to be misted regularly. But beware, you will kill a rosemary if you over-water it, underwater it, or winter it in a room with a woodstove. Wood heat is too drying for this dew-loving plant.

Rosemary seeds have a naturally low germination rate. Cuttings are the easiest way to propagate rosemary. Snip a 6–8–inch cutting and place 3 inches of the stem in a glass of water in a sunny window. It will root over a few weeks. Transplant into a pot with organic potting soil and place on a sunny deck for the first year. Rosemary loves lots of sun and prefers dryish soil. Keep protected from cold wind, as it can kill rosemary plants, small or large.

Rosemary prefers to come out of its pot and be placed directly into the ground during the growing season. I am careful when I decide to bring the rosemary plants outside in spring. It's usually the third week of April. For 1–2 weeks I keep them in their pots, protected from the wind against the south-facing wall of our farmhouse. After acclimating them to the out-of-doors, we wheelbarrow them into the garden and replant them in the same spots as the year before. We re-dig their holes, add a few shovels of compost, carefully tease them out of their pots, and gently and skillfully move them into their holes. We water each plant well with a hose in order to completely moisten the root ball before covering with garden soil. I am careful to use my feet to firmly tamp the soil in around each plant, leaving no air pockets.

Collecting: I learned to be bold when pruning rosemary plants from an older Italian farmer who kept a huge rosemary plant growing in the ground in his greenhouse in Massachusetts. His plant was over 5 feet tall and at least 6–7 feet wide. Very impressive. I collect several pounds of rosemary in the spring before bringing the plants outside and again in fall before bringing the plants inside. Each cut I make is at a juncture that will encourage new shoots to grow. Rosemary becomes bushier by boldly cutting 3–8-inch-long branches. Our spring and fall harvests are made into fresh tinctures and infused medicinal oils and dried for blending into a healing sitz bath. Throughout the rest of the year I snip smaller pieces for tea, soups, and cookies.

Juliette de Bairacli Levy wrote in her book *Common Herbs for Natural Health,* "As an herbalist, if my name could be associated with any plant I would choose rosemary. I use it more than any other plant and I love it most of all" (Levy, 1974, p. 123).

Actions: antibacterial, antioxidant, anti-inflammatory, antiseptic, antiviral, bitter tonic, brain tonic, bronchodilator, carminative, circulatory tonic, decongestant, heart tonic, nervine.

INDICATIONS

Circulation: Rosemary improves circulation and protects blood vessels. For people with spider veins and varicose veins, rosemary combined with hawthorn berries and gotu kola leaves can be helpful. Rosemary is beneficial for people with cold hands and feet. Take orally as a tincture or tea and use the pure essential oil in foot, hand, and full body baths and massage oils.

Digestion: Rosemary is warming and stimulating to the digestive tract and eases nausea and gas. The bitter and pungent qualities of rosemary improve the liver and gallbladder's ability to digest fats, stimulate digestion and appetite, and move food and waste products more effectively through the digestive system. Anne McIntyre says that rosemary raises digestive fire (agni), clears toxins from the body, and protects the liver from damage from chemicals and alcohol (McIntyre and Boudin, 2012, p. 111).

Hair: I rub a few drops of a pure essential oil of rosemary into my hair every morning to both awaken my mind and improve the quality of my hair. Rosemary tea, essential oil, and vinegar, when used topically, stimulate hair growth and eliminate dandruff. Several years ago ethnobotanist James Duke told me he believed that rubbing a few drops of rosemary essential oil into the scalp each day could be helpful in preventing cancer. He said the essential oil crosses the blood-brain barrier and enhances free radical scavenging activity in the body.

Immunity: Rosemary's warming and stimulating volatile oils contain antibacterial and antiviral properties that help clear excess mucous from the sinuses and chest and relieve colds, intestinal viruses, and bronchitis. Its antioxidant properties may reduce the risk of cancers, arteriosclerosis, and other oxidative diseases (Kuhn and Winston, 2001, p. 283).

Mind: Rosemary tea and tincture can be used to strengthen the mind, enhance alertness and memory, heighten concentration, improve cerebral circulation, reduce mental fog and mild depression, and ease vasoconstrictive headaches. For memory improvement and mental clarity I combine tinctures of fresh rosemary, bacopa, gotu kola, and sacred basil. Rudolf Steiner recommended placing rosemary oil in the bath for stimulating higher conscious faculties.

Musculoskeletal: Soaking the body in a warm bath infused with rosemary tea or a few drops of a pure rosemary essential oil is deeply relaxing and soothing to sore, painful muscles. Soaking tired feet in a warm rosemary foot bath also feels rejuvenating and relaxing. When chilled from working in the rain or being outside on a cold day, take a warm rosemary bath to prevent the onset of a cold or flu.

Respiratory: Rosemary warms the respiratory tract, helps prevent chest infections, and clears excess mucous from the sinuses and lungs. Its aromatic and antiseptic properties are especially beneficial during the cold, damp seasons of fall, winter, and early spring. I use it fresh in tea, in soups, in foot baths, full body baths, sinus steams, and as a massage oil for the chest.

PREPARATION AND DOSAGE

Aromatherapy: Place 1–2 drops of pure essential oil into a hot pan of steaming water. Place a towel over your head and pot and breathe in the medicated steam. Regular inhalations with rosemary essential oil or rosemary tea break up congestion in the sinuses and lungs and prevent or heal a chest infection. For a foot bath place 1–2 drops of the essential oil into a pan of steaming water and relax. For a full body bath, especially after a long day of gardening or during the cold winter months, place 4–5 drops of the essential oil into your bath tub, light a candle, and relax. If you find rosemary to be too stimulating at night, then use rosemary in the morning and use lavender, lemon balm, or rose essential oil in an evening bath.

Infusion: Add a few fresh sprigs to a cup of cool water and slowly heat in a pot with a lid until almost simmering and then turn off the heat and infuse for 5–10 minutes. Add 1 teaspoon dried leaves to 8 oz of hot steaming water, cover with a lid, and infuse for 5–10 minutes. Drink 1–3 cups per day. I especially love fresh ginger root tea infused with a few sprigs of fresh rosemary, a squeeze of lemon, and a spoonful of raw honey. I wait until my tea is cool enough to drink before adding the honey, as high heat destroys honey's beneficial enzymes.

Infused medicinal oil: Place finely cut fresh rosemary sprigs or dried leaves into a pint or quart glass jar. Fill the jar completely with organic olive, almond, or sunflower oil and cover with unbleached cheesecloth. Infuse at 80°F for 2 weeks. Strain off the oil and store the oil in a glass jar inside a dark, cool cupboard. This makes a lovely massage oil for sore, tired muscles, joints, and feet, or a rub for chest infections.

Soups: Make a strong soup stock with a large handful of fresh or dried rosemary leaves. Cover with water, place a lid on the pot, and slowly bring to a simmer. Turn off the heat before it boils and let infuse for 30–60 minutes. Rosemary is my favorite herb for adding into a nettle soup.

Tincture: (1:4) Finely chopped fresh or dried leaves. Use ¼–½ teaspoon, 2–4 times per day.

Safety considerations: Avoid during pregnancy.

Long term safety: safe

"I believe the most important steps for self-healing are:
To stop in our tracks, be still, take pause, and allow time to breathe and reflect.
The first cosmic law of ahimsa, practicing non-harm, informs that to fulfill our
humanity as a whole person, we must strengthen our capacity to gain awareness."
(TIWARI, 2011, PREFACE)

THE MOST DELICIOUS ROSEMARY COOKIES

1 stick organic butter or 8 ounces organic ghee
1 cup maple syrup
3 cups organic brown rice or spelt flour
4–5 heaping tablespoons of finely chopped fresh rosemary
Preheat oven to 350°F. Cream the softened butter or ghee with the maple syrup. Mix the chopped rosemary with the flour and add into the butter and maple syrup mixture. Spoon onto un-greased cookie sheets. Bake for 30 minutes or until lightly browned on the bottom. Enjoy.

SACRED BASIL

Latin name: *Ocimum sanctum*

Common name: Holy basil

Sanskrit name: Tulsi, Tulasi

Family name: Lamiaceae

Place of origin: India

Part used: leaf and flower

Energy: warming, dry

Flavor: bitter, pungent

When I first saw the seeds of sacred basil offered in a seed catalog, I immediately ordered them. I was drawn to this plant because of its name. Long revered in India for its abilities to purify the body, mind, and spirit and provide divine protection for the household, sacred basil is used to open the heart and mind and to enhance feelings of love and devotion. Many Indian households keep a Tulsi plant in a special clay vessel in their courtyard for its purification influence. Hindu worshippers of the god Vishnu often place a Tulsi leaf on their tongue during their morning prayers. Soon after I began growing sacred basil, well-known ethnologist, writer, and photographer Stephen Huyler stopped by Avena to purchase some herbs prior to traveling to India. We began a conversation about his work in India. Stephen gifted me with his book titled *Gifts of the Earth,* which includes a chapter about Tulsi.

Cultivating: We plant over 600 plugs (50 plugs per tray) with sacred basil seeds in the greenhouse in mid-April. The seedlings germinate within one week and begin to grow quickly thanks to longer days and warmer temperatures. In April the greenhouse temperature is kept between 65°F and 70°F. We transplant the seedlings into 3-inch pots in early May. Once there is no longer any chance of frost—usually early June— we trim the seedlings (making fresh tea from the trimmings) and transplant them into a hot sunny location in the garden that has been prepared with compost and mulched with straw. Over the first 2 weeks I keep a special eye on the sacred basil seedlings, watering them if there is no rain. I use a 3-year rotation for the garden beds that sacred basil is planted into. Garden maps with rotation plans help me remember where different annuals and perennials have been in previous years. Crop rotations are important for the long-term health of the soil and vitality of the plants.

Collecting: Sacred basil is one of my most favorite herbs. Before collecting I sit directly on the ground near a bed of sacred basil plants, close my eyes, and tune in to the spirit of Tulsi. I often chant quietly while harvesting the leaves and flowers. Regular cutting—once or twice a week—encourages the plants to become bushier and the tops to produce small flowers. Many pollinators value this plant, which is why I encourage its flowering. As with any basil, a light frost will blacken their tops, so be sure to harvest your plants when a frost is predicted.

Actions: adaptogen, antibacterial, antidepressant, antioxidant, aromatic, carminative, diaphoretic, expectorant, heart tonic, immuno-modulator, nervine, rejuvenative

INDICATIONS

Digestion: This special aromatic herb moves gas, lessens heartburn, and reduces fermentation and stress in the digestive tract. Its warming and stimulating qualities improve appetite and agni. Sacred basil lowers blood sugar levels, cholesterol, and triglyceride levels (Simon and Chopra, 2000, p. 160). It's a lovely tea to drink throughout the year to support good digestion and to awaken one's heart and mind to the beauty of each new day.

Immunity: Sacred basil reduces fever and flu symptoms, sore throats, coughs, and histamine-induced allergies (McIntyre and Boudin, 2012, p. 213). It enhances overall immunity and resiliency and protects healthy cells from the toxicity associated with radiation and chemotherapy (Simon and Chopra, 2000, p. 159). Sacred basil helps repair damage to the DNA from overexposure to radiation and protects the body against chemical carcinogens.

Lungs: Clears dampness and toxic ama from the lungs and upper respiratory tract. Human clinical trials have shown that sacred basil can increase vital lung capacity and reduce labored breathing, specifically enhancing prana and the vital life force (Pole, 2006, p. 280).

Mind: Sacred basil heightens awareness and mental clarity and enhances cerebral circulation and memory. It's beneficial for people with cloudy thinking, poor memory, lethargy, attention deficit disorder (ADD), and attention deficit hyperactivity disorder (ADHD).

Nerves: Sacred basil reduces levels of stress hormones, including cortisol. It's an important herb for building resiliency and enhancing a feeling of balance and well-being. Herbalist and Ayurvedic practitioner Anne McIntyre uses sacred basil for easing anxiety, depression, insomnia, and other stress-related conditions such as headaches and irritable bowel syndrome (McIntyre and Boudin, 2012, p. 213). Tulsi can be added to a tea with other nervine herbs such as green milky oat seeds, lemon balm, gotu kola, and rose petals to nourish and rebuild the nervous system and uplift the spirit.

PREPARATION AND DOSAGE

Aromatherapy: The pure essential oil of sacred basil clears the mind, opens the heart, and uplifts the spirit. A few drops of the essential oil, or an infused oil, can be gently massaged over the heart or on the back of the neck for protection. This is a very special plant to grow in your garden and in pots near the entrance to your home. Sit near these plants, on a daily basis, and deeply breath in Tulsi's healing fragrance through your nostrils. Let the aroma and energy of this sacred plant heal you.

Bath: Fresh sacred basil leaves and flowers are lovely to float in a warm bath or to use as part of a spiritual bathing ceremony. A strong infusion can also be made from the fresh or dried herb and added to bath water. A few drops of the pure essential oil of Tulsi can be placed into a foot or full body bath.

Spiritual baths: Different cultures combine aromatic herbs and flowers with water for spiritual healing ceremonies. Calendula, rose petals, rosemary, and sacred basil are herbs I use in spiritual baths. My deepest thanks to Rocio Alarcon from Ecuador for sharing this ceremony.

Ghee: Sacred basil can be mixed with ghee. Anne McIntyre writes: "Medicated ghees are used to enhance absorption of herbs, to nourish the nerves, and calm the mind. As ghee is easily absorbed it is a good vehicle to carry remedies deep into the tissues. One part of herb is simmered in four parts of ghee and sixteen parts of water until the water is evaporated. To test when it is ready, dip a piece of paper into the ghee and then put it into a candle flame. It should not crackle when it is put in the flame." (McIntyre and Boudin, 2012, p. 344)

Glycerite: (1:4) Fresh or dried herb, ¼–½ teaspoon, 3–5 times per day.

Infusion: Cover a handful of fresh leaves and flowers with cool water and slowly bring to a simmer, turn off the heat and steep covered, 10–30 minutes. For dried leaves and flowers, add 1 tablespoon to 8 ounces of hot water, steep covered, 5–15 minutes. For acute conditions drink 4 ounces of tea every 2 hours.

As a restorative tea, drink 2–3 cups per day. During the summer two large clay pots of tulsi sit on the deck at Avena Botanicals. Each morning I collect a handful of fresh leaves and flowers and place them in my glass quart water bottle. I sip on this sun infusion throughout the day. Tulsi's tiny purple flowers add beauty to the water and impart a mildly sweet and pungent flavor.

Infused medicinal oil: Fresh or freshly dried leaves and flowers can be infused in an organic carrier oil such as olive, sesame, or almond for two weeks at 100°F. This makes a special oil for foot and whole body massage, for anointing your third eye and heart chakra, and for use in Ayurvedic body treatments.

Powder: ½–1 teaspoon stirred in 4 ounces of hot water, 1–2 times per day. Grind your own or check with the herb supplier on the date the herb was powdered.

Tincture: (1:4) Fresh or dried herb, ¼–½ teaspoon, 3–5 times per day.

Safety considerations: Avoid during pregnancy. Combine sacred basil with cooling herbs such as lemon balm, yarrow, and elder flowers for a person with a high fever and respiratory infection. Consult with your health care provider if using anticoagulants.

Long-term safety: safe

The universe is a continuous web.
Touch it at any point and the whole web quivers.
—STANLEY KUNITZ

CALENDULA

Latin name: *Calendula officinalis*

Common names: pot marigold, Mary's gold

Family: Asteraceae

Place of origin: southern Europe

Part used: flowers

Energy: warm, dry

Flavor: bitter, slightly pungent

Cultivating: This brightly colored flower grows easily as an annual. For years I have grown an orange variety from Horizon Herbs. Calendula can be started inside or direct seeded into the garden in early spring and can also be grown in containers on a sunny deck. Calendula grows best in full sun and in well-drained soil that has compost worked into it. We start 1,000 calendula plugs inside the greenhouse near the middle of April and transplant them into the garden in mid- to late May. We seed another 1,000 plugs in mid- May and a third seeding of 1,000 plugs the first week of June.

We have created a 3-year rotation system in the garden for our calendula beds. I believe it's better for the soil and for the health of an annual crop to not be planted in the same bed year after year. The calendula seedlings we transplanted into the garden in late May usually diminish in size and vigor by early August and are pulled out of the ground and placed into the compost pile. I then direct seed either borage or oats into these empty beds. Our second and third plantings of calendula will flower into October if they have received enough water. We aim to pull our annual crops out of the ground by late October. Each bed is then covered with a thin layer of compost followed by straw. These composted and mulched beds are ready for us to plant a different annual into the following spring.

Collecting: Our first calendula planting begins to flower in late June. We collect the blossoms two to three times per week. I prefer to use my fingers when picking each flower. I have found finger-picking to be more efficient than using the Joyce Chen scissors. I also love feeling the flowers' sticky resin on my fingers. I am careful to not place enormous amounts of flowers into one basket as they will start to wilt if piled too high on top of each other. Once the harvest is complete, we carry our baskets into the herb drying room, record their weight, and lay them onto screens to dry. On average we harvest 300–350 pounds of fresh calendula each growing season. Calendula is an important ingredient in many of Avena's teas, salves, and face creams. We tincture several gallons of fresh flowers and also create a fresh succus (medicinal juice) by grinding both flowers and leaves together in a small amount of alcohol. The liquid succus is used as a spray for insect bites and for lessening scarring after surgery.

Actions: antiseptic, antifungal, anti-inflammatory, antimicrobial, diaphoretic, emmenagogue, lymphatic decongestant, immune tonic, vulnerary. Useful in vaginal suppositories and sitz baths.

INDICATIONS

Breasts: Calendula can be used topically and orally to reduce breast cysts, to ease tender, swollen breasts, and to improve lymphatic circulation. Beneficial when taken orally for women with post-mastectomy lymphedema and pain. Calendula also enhances the healing process when used topically for surgical wounds (Winston and Kuhn, 2001, p. 80). Calendula oil and salve are safe and effective for nursing women for healing sore and cracked nipples.

Digestion: Combine with other herbs for healing chronic colitis and gastric ulcers. Commonly used by herbalists and homeopaths as a mouthwash for healing gum tissue after tooth extraction and for healing the mucosa lining of a person's mouth that is inflamed or raw. The bitter and pungent taste of calendula warms the digestive tract and helps clear toxins from the gut (McIntyre and Boudin, 2012, p. 34). Beneficial in a sitz bath and salve for healing hemorrhoids and anal fissures.

Eyes: Calendula tea makes a soothing eyewash for tired, red eyes and for reducing inflammation of the eyelids from conjunctivitis and styes.

Gynecology: Calendula is an important addition to a formula for reducing fibroids and ovarian cysts (McIntyre, 2002, p. 77). Its antifungal, antiseptic, and anti-inflammatory actions make this a valuable herb to include in suppositories for healing vaginal infections and vaginal tissue that has been traumatized. Calendula can be used as part of a healing protocol for herpes and thrush. I always add calendula into a tea blend for a woman who has experienced sexual abuse. These flowers assist in the healing of both physical and emotional pain.

Immunity: Mildly stimulates the immune system. Add into teas or tinctures for reducing colds, flus, fevers, and swollen glands.

Liver: Calendula's bitter flavor stimulates the actions of the liver and gallbladder, increases bile secretions, and improves bowel function. Its ability to improve liver function aids the body in eliminating toxins.

Lymph: Calendula taken as a tea or tincture reduces lymphatic congestion and swollen glands.

Skin: Calendula promotes tissue repair, enhances the healing process of surgical wounds, softens and lessens scar tissue, reduces inflammation, and is helpful for slow-to-heal wounds when used topically as an oil, salve, diluted tincture, or succus. Calendula oil and salve are safe to apply on the perineum before and after birth. Use a salve to prevent scarring from chicken pox.

PREPARATION AND DOSAGE

Infusion: 1 tablespoon of dried flowers per 8 ounces of water. Steep in hot water for 10–15 minutes. Calendula is a favorite herb of mine to add into various tea blends for the joyful and bright color it imparts to a pot of tea in winter. Fresh flowers can be used throughout the summer months in sun teas or placed in cool water with other herbs and slowly heated to a simmer. Turn off heat and let infuse for 10–15 minutes.

Infused medicinal oil or salve: I have found that infusing dried calendula flowers into an organic carrier oil for 2 weeks makes a deeper amber-colored oil than when infusing fresh flowers. Use topically as often as needed.

Sitz bath: Combine calendula with herbs such as comfrey leaf, yarrow, rosemary, and lavender in a strong tea and use as a sitz bath for soothing inflamed perineal tissue or tissue around the opening to the vagina and for healing hemorrhoids.

Succus: Spray as often as needed for healing cuts, wounds, abrasions, and insect bites.

Tincture: (1:5) Fresh or dried flowers, ¼–½ teaspoon, 3–4 times per day.

Safety considerations: Avoid oral use during pregnancy. Safe for topical use during pregnancy. Be sure to wash a wound well before applying calendula, as it promotes rapid tissue healing. People with allergies to plants in the Asteracea family may need to avoid oral and topical use of calendula.

Long-term safety: safe

Flower essences

In Avena's garden we create flower essences by placing flowers in a small glass bowl filled with spring water. I prefer to collect the flowers early in the morning when flowers feel most alive. We place the bowl in a protected place in the garden for a few hours and let the sun's light infuse the flowers' healing qualities into the water. Flower essences act as catalysts to reawaken the natural life force within us. They support one's ability to transform limiting emotions and attitudes into more life-affirming ways of being.

Around the world herbalists and flower essence practitioners make flower essences. Instead of including information about the flower essence qualities of the flowers in this section, I wish to encourage you to directly connect with flowers, wherever you live, and observe, listen, and feel what the flower is communicating with you. Trust your intuition. There are wonderful books and flower essences to purchase to support your curiosity and healing process. However, your own direct experience with flowers will be your greatest guide.

LADY'S MANTLE

Latin name: *Alchemilla vulgaris*

Common names: bear's paw, lion's foot, dew cup

Family: Rosaceae

Place of origin: northern Europe

Parts used: flower, leaf, and root

Energy: cool, dry

Taste: bitter, astringent

Cultivating: This hardy and beautiful perennial grows well in full sun or partial shade. It creates a lovely hedge when massed in along a pathway. Lady's mantle also adds a magical touch to an herb garden when grown as a single plant. Every 4–5 years we divide our lady's mantle plants using a sharp, long-handled spade. We replant them with compost or pot them up and pass them on to other gardeners. Avena's lady's mantle beds were started from a few plants that were given to me over 25 years ago. Today more than 50 plants happily grow in a partially shady section of Avena's garden. Sometimes happy plants will reseed themselves, so keep an eye out for their tiny seedlings in the springtime. This plant is easy to propagate from root divisions.

Collecting: We collect the leaves and flowers early in the morning when dew-like droplets outline the edges of the leaves and a water droplet rests in the center of each leaf. Years ago herbalist Adele Dawson told me that the plant itself exudes these droplets and that the old alchemists believed these water droplets contained subtle healing qualities. Adele's teaching is what inspired me to gather and tincture this plant early in the morning when it is covered with dew.

We wait until midmorning when the flowers and leaves are free of the morning dew before collecting and laying them on screens to dry. Once they're dry, we mix them into tea blends. The delicate yellow flowers add a magical touch to tea.

Actions: astringent, diuretic, emmenagogue, regenerative tonic, styptic, uterine stimulant, vulnerary

INDICATIONS

Digestion: Useful as an astringent for acute diarrhea and as a mouthwash for soothing mouth ulcers. The salicylic acid found in the leaves helps reduce inflammation in the digestive tract.

Gynecology: Lady's mantle is a special herb with many healing benefits for women. As a uterine astringent it helps reduce heavy menstrual and menopausal bleeding. I often combine it with yarrow and a fresh shepherd's purse tincture for lessening menstrual bleeding. Lady's mantle tea and tincture eases emotional changes following a birth, after miscarriage, and during menopause. It helps tighten ligaments attached to the uterus, making it a beneficial herb for women after giving birth. It can be combined with raspberry leaf tea and the tinctures of astragalus root, burdock root, and true Solomon's seal root for supporting a woman with a prolapsed uterus. Australian herbalist Ruth Trickey recommends lady's mantle for adolescent and perimenopausal women experiencing heavy bleeding that is associated with hormonal fluctuations. Anne McIntrye recommends this herb for women experiencing pelvic congestion and heavy bleeding associated with fibroids or endometriosis.

For supporting conception, lady's mantle can be combined with teas such as raspberry leaf, red clover, oat seed, and rose petals; root powders of ashwagandha and shatavari; tinctures of ginger, chaste tree, and dong quai; and flower essences specific for each individual woman. Once pregnancy occurs, discontinue these herbs and consult with an herbalist and midwife on the appropriate herbs and foods to use.

Lady's mantle tea is beneficial as a gentle healing douche or wash for mild cervical erosion and for reducing excessive vaginal discharge. It's a valuable herb for women who have experienced stress or trauma to the uterus from a miscarriage, abortion, difficult labor, pelvic inflammatory disease (PID), surgery, use of an IUD, or sexual abuse. I combine lady's mantle with rose petals, calendula, self-heal, and white yarrow as a tea, tincture, and bath. Herbs are important healing allies for women who have experienced any kind of trauma.

Nerves: Years ago Adele Dawson taught me to use lady's mantle as a tea, tincture, or flower essence for women in any kind of a major life transition such as pregnancy, menopause, entering or finishing school, moving, birthing a creative project, or beginning or ending a relationship. Adele pointed out that the cloaklike leaves of lady's mantle offer energetic protection and strength to women. Adele also taught me to collect the large drop of dew that sits in the center of each leaf and place it in spring water. Add a teaspoon of brandy to the spring water to preserve the dew essence. This magical dewdrop essence can be healing to women who have experienced trauma. Juliette de Bairacli Levy wrote: "The true name of this plant is Arabic—alkemelych, from the alchemists who were the ancestors of the chemists, and its name shows the high esteem in which the Arabs held—and still hold—this herb" (Levy, 1974, p. 89). The old alchemists used the dew droplets, which the plant itself exudes, as part of their formulas for longevity.

Skin: Adele also encouraged me to place my face in the center of a large dew-laden lady's mantle leaf at dawn. This is one of my favorite early morning rituals. A tea, lotion, or crème made with lady's mantle flowers and leaves helps heal wounds, sores, and rashes.

Urinary tract: Lady's mantle's astringent properties are helpful for bladder and rectal prolapse and urinary incontinence. It can be combined with yarrow, witch hazel, rosemary, comfrey leaves, and calendula flowers as a sitz bath for healing anal fissures, episiotomy incisions, and hemorrhoids.

PREPARATION AND DOSAGE

Infusion: Cover a handful of fresh leaves and flowers with cool water and slowly bring to a simmer, steep covered, 10–30 minutes. For dried leaf and flowers, add 2–3 teaspoons to 8 ounces of hot water, steep covered, 5–15 minutes.

Tincture: (1:3) Fresh flowers and leaves, ¼–½ teaspoon, 3–5 times per day.

Dewdrop essence: Collect the early morning dew drops using a clean dropper. Fill a 1-ounce bottle and store in the refrigerator. Place a few drops directly under the tongue as needed.

Safety considerations: Avoid during pregnancy. Anne McIntyre recommends using lady's mantle in the last 10 days of pregnancy to prevent excessive postpartum bleeding (McIntyre and Boudin, 2012, p. 13).

Long-term safety: safe

"I really believe that to stay home, to learn the names of things, to realize who we live among . . . The notion that we can extend our sense of community, our idea of community, to include all life forms — plants, animals, rocks, rivers and human beings — then I believe a politics of place emerges where we are deeply accountable to our communities, to our neighborhoods, to our home. Otherwise, who is there to chart the changes? If we are not home, if we are not rooted deeply in place, making that commitment to dig in and stay put . . . if we don't know the names of things... Then we enter a place of true desolation."
—"The Politics of Place: An Interview with Terry Tempest Williams" by Scott London

LAVENDER

Latin name: *Lavendula officinalis*

Common name: lavender

Family: Lamiaceae

Place of origin: native to southern Europe

Part used: blossoms, leaves

Energy: cool, dry

Flavor: bitter

Cultivating: This woody perennial needs full sun and a warm and protected place, especially in Maine. Prefers sandy and well-drained soil. Two 40-foot-long hedgerows of lavender grow in one of the driest areas of our farm. I call this garden the Mediterranean garden. It's here that lavender, thyme, hyssop, and blessed thistle thrive. Lavender can be started indoors from good-quality seed if you have the time and patience. If you are like me—eager for a blooming lavender hedge—then look for a local perennial gardener who sells their own hardy, field-dug lavender plants. They can be transplanted into the ground starting in late April. Be sure to tamp the soil firmly around each plant-leaving no air pockets so they don't heave during winter thaws. In zones colder than 5, mulching lavender and thyme with straw or balsam boughs gives them extra protection. Lavenders are hardy to -20F.

Collecting: Honey bees adore the nectar of lavender blossoms—they're the first to sip from these delicate flowers. We carefully cut the lavender blossoms soon after they open, using either Joyce Chen scissors or Felco #310 harvest shears, and place them into flat baskets. The blossoms are quickly made into tinctures, non-alcohol glycerites, and laid out or hung to dry for wintertime teas. I like adding fresh lavender blossoms to honey and dried blossoms to shortbread cookies. (See recipe below.)

Actions: antiseptic, anodyne, anti-inflammatory, antispasmodic, carminative, emmenagogue, expectorant, nervine

INDICATIONS

Digestive: Eases gas and nausea and relaxes the digestive tract. Anne McIntyre recommends using lavender for stomach and bowel infections that cause vomiting and diarrhea and for healing inflamed gums and mouth sores. Lavender's antiseptic oils help eliminate unwanted bacteria from the colon. Lavender combined with fennel seed is effective for healing gas and indigestion.

Gynecology: A favorite herb of mine to combine with green, milky oat seeds, rose petals, blue vervain, and skullcap for easing acute premenstrual and menopausal stress and agitation. Lavender can be added into a sitz bath blend for soothing and healing the perineal tissue after birth.

Mind: Eases stress-induced headaches, relaxes and calms the mind, reduces mental and emotional agitation, and lessens feelings of unrest and depletion. As a gentle nerve stimulant, lavender can help transform mild depression and seasonal affective disorder (SAD). I combine lavender with lemon balm, St. John's wort, and calendula blossoms for people with seasonal affective disorder (SAD).

Nerves: Lavender eases stress, anxiety, insomnia, low self-esteem, and nervous tension and restores vitality to a person experiencing nervous exhaustion. Lavender's aromatic oils help uplift a person's spirit, shift inner disharmony, and soothe and comfort the heart.

Respiratory: An excellent herb for resolving respiratory infections such as bronchitis and pneumonia and for soothing sore throats, tonsillitis, and laryngitis. Years ago I successfully healed a tough respiratory infection (borderline pneumonia) using a combination of lavender, licorice root, thyme, and elecampane root tinctures.

Skin: Lavender's cooling, anti-inflammatory and antimicrobial properties are very effective when used topically for healing burns, infected cuts, wounds, insect bites, and red and irritated conditions such as acne and eczema.

PREPARATION AND DOSAGE

Aromatherapy: The fragrance of lavender soothes and comforts a person who feels tired and stressed. I place a few drops of an organic essential oil of lavender on a piece of cotton or tissue and place inside my pillow case when my sleep is disrupted from traveling or I am unable to fall asleep because of stress. A little vial of lavender essential oil is part of my home first aid kit and my travel first aid kit. I use it topically, undiluted, for burns, cuts, wounds, and stress headaches and for reducing scar tissue. A few drops of the essential oil placed in a pot of hot steaming water creates an antiseptic and anti-inflammatory steam that is beneficial for people with congested sinuses or a respiratory infection. Place a towel over your head and the pot and breathe in the healing aroma for 3–5 minutes, several times a day. A few drops of the essential oil can be placed into a warm foot or full body bath or shower for calming the mind and reducing stress.

Glycerite: (1:4) ¼–½ teaspoon, 3–4 times per day

Infusion: Fresh or dried blossoms can be used in tea, 1 teaspoon per 8 ounces of water. Slowly bring water to a simmer and turn off the heat. Infuse, covered, for 10–20 minutes. The longer it infuses, the more bitter the tea tastes. Drink 1–2 cups per day. A strong lavender infusion is a lovely addition to a foot or full body bath.

Infused oil: Infuse fresh or dried blossoms in an organic carrier oil such as apricot kernel, almond, or olive at 100°F for 2 weeks. Strain and store in a glass bottle. Use topically as a massage oil.

Sitz bath: Combine lavender with herbs such as calendula blossoms, comfrey leaf, rosemary, and yarrow flowers. Make a strong infusion by using fresh or dried herbs. Strain and use as a sitz bath for postpartum inflammation and for healing hemorrhoids.

Tincture: (1:4) Fresh or dried blossoms, ¼–½ teaspoon, 3–4 times per day

Safety considerations: Avoid using orally during pregnancy, as lavender is a uterine stimulant.

Long term safety: safe

Lavender Shortbread Cookies

    1 cup whole spelt or whole wheat flour
    1 cup white spelt flour
    1 cup brown or white rice flour
    1 tsp non-aluminum baking powder
    ½ cup organic ghee, butter, or sunflower oil
    2/3 cup pure maple syrup
    2 tsp pure vanilla
    2–3 Tbl dried culinary lavender or 2 Tbl finely chopped fresh lavender flowers
    Preheat oven to 350°F. Use ghee or oil to grease your cookie sheets or a shortbread mold. Mix together the dry ingredients and wet ingredients in separate bowls. Add wet ingredients to dry and mix together to form a smooth dough. Add flour as needed to ensure the dough is smooth. Place the dough between waxed paper and roll it out to ½-inch thickness. Take off the top sheet of waxed paper and either form individual cookies or invert the dough into the mold. If using the mold be sure to press the dough firmly into the mold. Bake for 30 minutes or until edges are slightly golden. Wait another 30 minutes before releasing the shortbread from the mold.

ROSE

Latin names: *Rosa rugosa, R. damascnea, R. centifolia, R. gallica*

Common name: rose

Family: Rosaceae

Place of origin: *R. rugosa* is originally from China. It has naturalized itself along the coast of Maine.

Part used: flowers (whole), petals, hips

Energy of flowers: cool and moist

Flavor: bitter, pungent, slightly sweet

Cultivating: This hardy perennial prefers full sun and well-drained soil. The easiest way to establish *Rosa rugosa* plants is to purchase them from a reputable organic rose grower or nursery. Once planted, be sure to mulch the roses with straw, seaweed, or organic buckwheat hulls. We use straw in our circular rose garden and seaweed and organic buckwheat hulls along our hedgerow of roses (organic buckwheat hulls are available through FEDCO). Buckwheat hulls work well as a mulch and are easy to apply early in the spring once the roses have been pruned. I learned to be bold when pruning *Rosa rugosa* from an experienced gardener and permaculturist, Lauren Cormier, who learned rose pruning from her mother. In early April we cut our shrubs back by half. This amount of pruning encourages the roses to produce an abundance of flowers.

Collecting: Avena's roses begin blooming in mid-June. Collecting rose petals early in the morning when the air is still is one of my favorite garden tasks. The ruby-throated hummingbirds are often darting about the rose garden at dawn. Our roses bloom continuously for almost 4 weeks. Every day there are hundreds of new blooms to pick. If there have been a few foggy or rainy days when we cannot collect roses, then I nip the spent blossoms to encourage the plants' further blooming. We prepare over 15 gallons of fresh rose petal elixir and dry as many petals as we can collect before the Japanese beetles emerge in July. (Instead of attacking the

beetles with a fearful, disgusted, or warlike mentality, I am attempting a biodynamic approach of ashing the beetles. Visit the Josephine Porter Institute's website [http://www.jpibiodynamics.org/] for more information on ashing "pests." There are very specific astronomical times for using the ashing method.)

I gather the rose hips in October after the first frost, cut them in half, and lay them out to dry on screens or in a food dehydrator. I enjoy rose hip tea, especially in winter, combined with fresh ginger root and red lycii berries.

Actions: antidepressant, anti-inflammatory, antimicrobial, aromatic, astringent, carminative, decongesting, detoxifying, nervine

INDICATIONS

Digestion: The antimicrobial activity in rose petals helps resolve infections in the gastrointestinal (GI) tract. Its astringent and cooling properties reduce hyperacidity, heartburn, enteritis, and diarrhea. Rose petal tea helps heal mouth and gut ulcers. Anne McIntyre recommends using rose petals to clear toxins from the gut.

Gynecology: The cooling and decongesting qualities of rose petals help relieve uterine pain and spasms, uterine congestion, and heavy menstrual bleeding, and clear vaginal infections and inflammation. Roses reduce hot flashes and premenstrual and menopausal stress, including feelings of low self-esteem. Roses ease irritability and anger and help lessen painful symptoms associated with menstruation, endometriosis, and fibroids. A pure rosewater spray cools hot flashes, PMS heat, emotional agitation, red and irritated skin eruptions, and inflamed vaginal tissue. Rose petal elixir is a special remedy for a woman seeking to understand and explore her sexuality.

Immunity: Dr. Vasant Lad, an Ayurvedic doctor and author, recommends macerating fresh rose petals in honey or raw sugar for sore throats and for healing mouth sores. Rose petal tea, tincture, and elixir help clear excess heat and toxins from the body and resolve cold and flu symptoms. Rose hip syrup has traditionally been used by herbalists to prevent infectious colds.

Nerves: Roses relax the nervous system and ease nervous depression, anxiety, agitation, and impatience. David Winston recommends rose petals for people experiencing emotional trauma (including trauma-induced depression) and posttraumatic stress disorder (PTSD). Roses soothe, calm, and comfort an unsettled and grieving heart. They gently open the heart and mind and inspire a feeling of love and compassion for oneself and others.

Skin: Rose petals cleanse and astringe the face; clear acne, skin blemishes, boils, abscesses, eczema, and psoriasis; and reduce inflamed eyelids (Pole, 2006, p. 252). Rosewater is my favorite herbal water to use after I cleanse and wash my face.

PREPARATION AND DOSAGE

Aromatherapy: A pure rose petal essential oil is uplifting and restoring to the heart and nervous system. The fragrance of rose helps lift depression; eases grief, sadness, and heart palpitations; and dispels physical and mental fatigue.

Glycerite: (1:3 or 1:5) I prefer to use fresh petals but dried can be used. Take ¼–½ teaspoon, 2–3 times per day.

Honey: Completely fill a sterilized glass pint or quart jar with fresh rose petals and cover with raw honey. Let infuse for a month in a sunny window. The petals can stay in the honey indefinitely.

Infusion: Place 1–2 tablespoons of fresh or dried rose petals in 8 ounces of cool water and slowly warm. Turn off the heat and let infuse for 10–15 minutes, covered. The longer the petals infuse, the more bitter they taste.

Tincture: (1:3 or 1:5) I prefer to use fresh petals but dried can be used; ¼–½ tsp, 2–3 times per day.

Rose Petal and Cardamom Cordial

1 cup fresh whole rose petals
1 tablespoon freshly chopped sacred basil leaves and flowers
1 teaspoon freshly ground cardamom seeds (not pods)
Place the herbs in a clean glass quart jar and cover the herbs completely with brandy, approximately 2 cups and a bit more. Place a non-BPA plastic lid on the jar, label, and place in a cool cupboard for 2-4 weeks. Strain off the herbs using a funnel and unbleached cheesecloth. I let the liquid drip through the cheesecloth. Though I am tempted to squeeze the cloth, I don't as squeezing fresh herbs creates a cloudy cordial. Sweeten with ½ cup of maple syrup, honey, or organic sugar.

SELF-HEAL

Latin name: *Prunella vulgaris*

Common names: all-heal, carpenter's herb, heart of the Earth

Family: Lamiaceae

Place of origin: Europe and Asia

Part used: flowering tops with leaves

Energy: cool, moist

Flavor: bitter, pungent

Cultivating: Self-heal is a non-native wildflower that has naturalized itself along wooded trails and in pasture lands and lawns in North America. Plants growing in wooded areas tend to be taller than plants found growing in full sun. Self-heal reseeds itself easily, especially in moist and shady areas. You can attempt to establish a bed in your garden by either collecting seeds from the wild and scattering them in a prepared, moist seed bed or planting purchased seeds (Horizon Herbs) in pots in the spring. I fill a 50-plug tray with organic potting soil and scatter the seeds on top of the soil. The seeds need light to germinate. I place the flat on the floor of the greenhouse underneath a potting bench where it's cool and damp yet still receives enough light. I gently water the seedling tray as needed. The seeds take 2–3 weeks to germinate. Plants become easily pot-bound and appreciate being transplanted into larger pots. Space plants 8–12 inches apart.

Many common weeds are herbalists' favorite healing plants, including burdock, dandelion, nettle, and self-heal. We want to be thanking them, eating them, and making medicine from them instead of haphazardly pulling them out of the garden or lawn and tossing them aside. Get on your hands and knees in spring and learn to identify the plants when they first emerge in spring. I never cease to be amazed and delighted by the magic I observe and feel when kneeling or lying with my belly upon the Earth.

Collecting: Collect the beautiful bluish-purple flowering tops soon after the flowers open and lay them on screens to dry. They're a lovely addition to any healing tea, fresh or dried. Self-heal is also a special herb to add into a healing bath. In Peg Schafer's book *The Chinese Medicinal Herb Farm,* she writes that for the traditional Chinese herb market self-heal is harvested when the majority of the flowers are starting to turn brown. Peg's book is a valuable resource for any herbalist or gardener interested in growing Chinese medicinal herbs.

Actions: astringent, antiseptic, anti-inflammatory, mild diuretic, lymphatic decongestant, vulnerary

INDICATIONS

Liver and kidneys: Self-heal is known by traditional Chinese herbalists to clear heat from the liver. The symptoms of heat in the liver may include red face, headaches, feeling overheated, and easily becoming frustrated and angry. Herbalist and author Matthew Woods writes:

> The late herbalist William LaSassier considered it (self-heal) a very important remedy for building and nourishing the kidneys. It improves diuresis, not by actively pumping the kidneys, but by strengthening them. He used it especially when there was dental decay combined with kidney trouble. This is perhaps a condition caused by inability to deal with protein properly, resulting in excess protein waste products (urates) burdening the liver and kidneys, precipitating with calcium to weaken teeth and bones. He preferred the traditional Chinese medicine explanation: the kidneys strengthen the bones and teeth.
> (Wood, 2008, p. 404)

Lymph: Anne McIntyre recommends using self-heal for swollen glands, mumps, glandular fevers, and mastitis and as an astringent herb for diarrhea and colitis. David Winston recommends combining self-heal with other lymphatic herbs (calendula, cleavers, red clover, red root, violet) for reducing lymphatic swelling in the breast tissue, for softening and reducing cysts and lipomas, easing sore throats, and resolving blood stagnation from repeated bruising.

Skin and wounds: An excellent wound-healing herb. The leaves can be chopped and used as a fresh poultice, and the whole plant can be used fresh or dried as an antiseptic wash for a wound or in a salve for soothing insect bites and stings. Self-heal's astringent properties help heal mouth ulcers when used as a mouthwash. Self-heal can be combined with other herbs in a tea, tincture, or salve for more quickly resolving a herpes outbreak. I recommend using self-heal with calendula, yarrow, and rose petals in a tea blend and/or sitz bath for a woman who has experienced any kind of sexual trauma. Women worldwide, every day, deal with way too many sexist and misogynist attitudes and comments. Self-heal's name reminds us that the body has its own ability to heal, both inwardly and outwardly.

PREPARATION AND DOSAGE

Infusion: Place a handful of fresh flowers and leaves in 1–2 cups of cool water. Slowly bring to a simmer, cover, and infuse for 5–10 minutes. Drink 1–3 cups as needed. Use 3 or more dried flowers per cup of hot water, alone or in combination with other herbs such as calendula, oat seed, rose petals, red clover blossoms, lady's mantle, and lemon balm. Steep for 10–20 minutes and drink 1–3 cups as needed.

Infused oil: Collect fresh flowers and leaves once the dew has dried and place them in a pint glass jar and cover with organic olive oil. Be sure to fill the jar completely with herb and oil, leaving 1–2 inches of headroom. Cover the top of the jar with unbleached cheesecloth and set in a warm window for 2 weeks to infuse. After 2 weeks, separate the oil from the plant matter by pouring it through the cheesecloth. Let the oil sit for a day in a glass jar with a lid. If water appears at the bottom of the jar, you'll need to carefully siphon off the oil into another glass jar. This oil can be used as is or made into a salve by adding beeswax.

Tincture: (1:4) I prefer to make the tincture with fresh flowers and leaves, gathered early in the morning. Use a few drops of the tincture, up to 1 teaspoon, as needed.

Topical: Use a tea of self-heal as an antiseptic wash, or use freshly chopped leaves as a poultice for healing cuts and wounds.

Safety considerations: safe

The famous English herbalist John Gerard (1545–1612) said that "the decoction of Prunell made with wine and water, doth join together and make whole and sound all wounds, both inward and outward." Another British herbalist, Nicholas Culpeper (1616–1654), said, "Self-heal whereby when you are hurt, you may heal yourself."

Self-heal Flower Tea

I enjoy growing, collecting, and drying flowers for winter and springtime teas. Maine winters can be long and flowers added into tea brighten my day. I usually dry and store a half-gallon jar full of self-heal flowers. Choosing which flowers and herbs I want for my daily tea totally depends on my mood. Most days I add a few self-heal blossoms into my tea blend as I believe they support the body's innate ability to heal.

YARROW

Latin name: *Achillea millefolium*

Common name: milfoil

Family: Asteraceae

Place of origin: Europe

Part used: flowering tops

Energy: neutral, dry

Flavor: bitter, pungent

Cultivating: Yarrow is an easy perennial to grow from seed. We start 400–600 plugs inside the greenhouse in late March by pressing the seeds into the soil. Seeds need light to germinate. We transplant the seedlings in mid-May into sunny garden beds that are well drained, spacing them 8–12 inches apart. Yarrow spreads quickly. After 3 years yarrow needs to be divided and moved to a new garden bed in order to revitalize the plant's vigor. The alternative to moving or tilling the plants is to sheet mulch the bed in fall with cardboard and a thick layer of straw. This sheet mulching practice allows the yarrow to decompose back into the soil (beneficial for the soil) and avoids using a tiller, which damages earthworms and soil structure.

Collecting: We use Felco #310 harvest shears when gathering the delicate white flowers. Yarrow's blossoms continue to open over a few weeks, so several harvests can occur if you're growing a large bed of yarrow. For tinctures I prefer to collect the flowers early in the morning when they're freshest and most vibrant. We wait for the dew to evaporate before collecting the flowers for drying.

Actions: antiseptic, anti-inflammatory, antispasmodic, antiviral, astringent, bitter tonic, decongestant, diaphoretic, hepatoprotective, styptic, vulnerary

INDICATIONS

Bladder: Yarrow is helpful in formulas for acute cystitis (with or without bleeding), irritable bladder, and incontinence (Wood, 2009, p. 55).

Cardiovascular: Some naturopathic doctors and European-trained herbalists use yarrow in combination with hawthorn flowers and berries, linden flowers, and motherwort for hypertension. Yarrow improves circulation and is often added into a formula for varicose veins.

Digestion: As a bitter tonic, yarrow stimulates appetite and improves digestion and assimilation. It helps to clear toxins from the gut and soothe inflammatory digestive problems. David Winston recommends using yarrow as part of a formula for conditions such as irritable bowel syndrome (especially with diarrhea or blood in the stool), leaky gut syndrome, gastric ulcers, mucus colitis, and diverticulitis.

Gynecology: I include yarrow in women's formulas for improving circulation to the pelvic area and easing menstrual cramps. As a uterine relaxant yarrow relieves pain, and as a uterine stimulant it increases muscular tone. For lessening heavy menstrual bleeding and reducing postpartum bleeding I combine yarrow with fresh shepherd's purse and lady's mantle tinctures. Yarrow helps reduce inflamed ovaries and hemorrhoids.

Immunity: Traditionally, yarrow has been combined with elder flowers, peppermint, and catnip (or catmint) and given as a hot infusion for lowering an acute fever. Hot yarrow tea stimulates sweating, which helps clear heat and toxins during a cold, fever, or flu. Juliette de Bairacli Levy used yarrow for all fevers, including intermittent ones. Matthew Wood recommends using yarrow for bringing out rashes associated with chicken pox or measles.

Skin and wounds: Yarrow's antiseptic and anti-inflammatory volatile oils and astringent properties are valuable for cleansing and healing open wounds and cuts, stopping bleeding, soothing rashes and hemorrhoids, and easing perineal pain after birth (in a sitz bath). I recommend drinking yarrow tea or taking a tincture for healing bruises (old and recent) and for people who bruise easily because their skin is "papery" thin and their circulation compromised. Adele Dawson recommended using yarrow in a salve or creme for healing skin irritations and for reducing the itching of dry scabs that cover a cut or wound.

Years ago I accidentally took a small chunk of skin off the top of one of my fingers while using a pair of Felco shears. I immediately chewed fresh yarrow leaves and placed them onto the wound to staunch the bleeding and took homeopathic arnica (30c), orally, to ease the pain. A few hours later, I went to show a homeopathic medical doctor my finger. Dr. Vandersloot was impressed with how yarrow had stopped the bleeding. Every night for 2 weeks I soaked my finger in a strong infusion of yarrow and calendula. During the day I kept it bandaged and moist with Avena's Heal-All Salve. My finger healed well, with almost no sign of scar tissue!

PREPARATION AND DOSAGE

Aromatherapy: A pure essential oil made from yarrow flowers is blue in color and contains strongly antiseptic and anti-inflammatory properties. This essential oil is a valuable remedy to keep in one's home medicine chest and in one's traveling first aid kit. It can be used undiluted on insect bites, scrapes and abrasions, and bruises and swellings.

Infusion: Place 1–2 teaspoons of dried herb into 8 ounces of hot water and cover. Steep for 15–20 minutes. Drink 1–4 hot cups of tea per day. When using fresh flowers, place a handful of the flowers into a glass or stainless steel cooking pot, cover the flowers with cool water, place a lid on the pot, and slowly heat to a simmer. Turn off heat and let steep for 15–20 minutes.

Tincture: (1:3) Fresh or dried flowers, ¼–½ teaspoon, 4–5 times per day as needed.

Infused oil: Infuse fresh or dried flowers into an organic carrier oil of your choice for 2 weeks at 100°F. Pour off the oil through unbleached cheesecloth and store in a glass jar in a cool, dark cupboard. This oil can also be made into a salve.

Safety considerations: Avoid during pregnancy. People with severe allergies to Asteracea Family plants like ragweed and chamomile may want to avoid using yarrow.

Long-term safety: safe

Herbal Support for Gardener's during Tick Season

Dress appropriately. Do thorough tick checks daily. Pay attention to your body and stop before you feel overworked or tired. We are more vulnerable to tick bites when tired or stressed. Care well for your immune, digestive and nervous system throughout the year. Consider a year-round daily meditation practice. Rest and restore yourself during winter. I use **Avena's astragalus tincture** or tea, 3-4 times a week throughout the winter and then on a daily basis starting a month before the gardening season begins all the way through fall. Astragalus improves the integrity of your energetic boundary and helps you be less "thin-skinned" and stronger in your ability to stand upright, both emotionally and physically (see pages 148-151). I also place 1-2 droppers of **Avena's fresh Schisandra berry tincture** into my quart water bottle every day during the gardening season to strengthen my immunity and liver and to ensure my energetic boundaries are vibrant and strong. Schisandra is one of my favorite restorative herb for enhancing resiliency and helping my spirit feel peaceful and settled (see pages 209-212). My other favorite restorative herb is **Avena's double-extracted Eleuthero tincture**. I use it 1-2 times/day throughout the gardening season. The taste is sweet and grounding and helps me adapt to stress with more ease and flexibility. **Japanese knotweed** is another herb to consider using on a daily basis throughout the gardening season so that the roots beneficial properties are in the bloodstream. It is strongly antibacterial with anti-spirochetal activity. At Avena, we take extra care to ensure the Japanese knotweed roots that our farmer friends dig for us are tested for heavy metal contamination before we prepare our tincture.

Refer to Wolf Storl's book *Healing Lymes Naturally* and any of Stephen Buhner's books on lymes and co-infections.

ELDER
Latin names: *Sambucus canadensis, S. nigra*

Common name: elder bloom

Family: Caprifoliaceae

Place of origin: North America, found growing in northern Quebec south to Florida

Part used: flowers, berries

Energy: Flowers are cooling, berries are warming.

Flavor of flowers: sweet

Flavor of berries: sweet, sour, spicy

Cultivating: Elderberries grow wild throughout Maine and the northeast. We cultivate them on our farm because their flowers are beautiful and attract pollinators. The elderberries' close proximity to our gardens allows us to easily gather the delicate flowers as soon as they open and the berries as soon as they turn purple. Elderberries prefer well-drained soil, which is high in organic matter. I have noticed that trees planted on our farm in wetter soil are less productive than trees planted in well-drained and well-composted soil. Elderberries will be most productive when planted in full sun and when they receive an inch of water per week during the summer months. These medicinal shrubs benefit from applications of compost early in the spring. After applying compost, we mulch our shrubs with ramial wood chips (the tops of deciduous trees that are no larger than 2 ½ inches). I highly recommend purchasing Michael Phillips's book *The Holistic Orchard: Tree Fruits and Berries the Biological Way* to better understand planting and tending elderberries and other kinds of fruits, along with excellent information about soil, ramial wood chips, and safe solutions to pest and disease challenges. Check out FEDCO's tree catalog (they ship) for good-quality fruit trees, including elderberries, and review the resources listed in Michael's book.

Elderberry cuttings: My long-time friend and organic fruit grower Liz Lauer has had great success propagating elder from cuttings in a rather unusual way. She prunes her branches after the leaf buds have broken dormancy but have not yet leafed out. She cuts the branches into several pieces, each with 3–4 nodes. She dips the lower part of the bud stick into water and then into a rooting hormone and buries 1–2 nodes in a soil medium in a pot, leaving 1–2 buds above soil level. She recommends dipping the tip of any non-terminal bud into melted beeswax to prevent it from drying out. Every branch has only one terminal bud at its top, which of course you would not dip into beeswax. Liz places the pots outside her greenhouse, keeps them watered throughout the summer, and plants them in the ground in autumn once they have gone dormant. Even a small-sized elder shrub will need a 4–5 foot wire fence to protect it from nibbling deer.

Collecting: I keep an eye on our elder shrubs during the later part of June for the first sign of their blossoms opening. Early in the morning—with Felco #310 shears and our wooden apple ladder—we begin gathering the elder flowers. Each is covered with tiny florets shaped as umbels (umbrella-like). We carefully place the umbels, laden with yellow pollen, into our baskets, which are lined with thin cotton cloth (cloth prevents the pollen from slipping through the small open spaces of baskets). We immediately tincture the flowers and pollen, some in organic alcohol and some in organic vegetable glycerin. The rest we lay to dry on a thin cotton cloth placed on our drying screens.

We collect the purple berries in early September using Felco #310 shears (by now you can tell that these Felco #310 shears are an extension of my hand). Cleaning the berries off their tiny stalks takes time, but I welcome this meditative task. We tincture the whole fresh berries in alcohol and vegetable glycerin. It makes a beautiful deep purple and delicious tasting elderberry elixir that both children and adults love. Just by chance, a naturopathic student in Australia tested Avena Botanicals Elderberry Elixir in the school's "lab" and found that our elixir contained the highest amount of antioxidants in all of the products they tested!

INDICATIONS FOR FLOWERS

Immunity: The flavonoid called kaempferol that is found in elderflowers inhibit estrogen-related cancers and contain anti-allergic, antibacterial, anti-inflammatory, and antiviral properties

Herbalist Maia Toll says: "In late summer and early fall, Elder presents her berries. This is her crone stage. Her young limbs have hardened and become less flexible. Her light flowers have matured into dark berries, white and black, yin and yang. While the flowers help cool the body, the berries are mildly warming. Use the berries for flus, inflammation along the nerves, and stubborn coughs. I think of the flowers when the cough is high in the lungs, the little 'hh, chh' that children often do. I turn to the berries when the cough is deep."

(Kuhn and Winston, 2001, p. 172). The flowers and berries are being extensively researched for their ability to inhibit viral infections from replicating in the body. Years ago herbalist Mary Bove taught me to use a hot infusion of elderflowers, peppermint, yarrow, and catnip for lowering fevers. Elderflowers are safe to give to children for relieving fevers, flu, colds, and eruptive diseases such as chicken pox and measles and can be combined with lemon balm. Elderflower and sage tea is an effective gargle for sore throats. I commonly give elderflowers to a child or adult with a blocked Eustachian tube and also recommend they receive a lymphatic massage to encourage lymphatic drainage. For lowering a fever that accompanies a flu I often combine elderflowers with elderberries and boneset.

Mouth: An infusion of elderflowers, combined with calendula and chamomile, can be used as a mouthwash for healing inflamed gums and mouth ulcers. David Winston recommends using elderflowers, calendula, chamomile, and sage as a mouthwash for inhibiting inflammation caused by periodontal pathogens.

Respiratory: Elderflowers have long been valued for their ability to relieve upper respiratory infections, including sinusitis, tonsillitis, laryngitis, and colds. Elderflower's relaxing qualities ease bronchial spasms and work well as a decongestant. A hot cup of elderflower tea with a pinch of chamomile and catmint helps relax a restless and agitated child or adult, especially at the onset of an illness, and supports a restful sleep.

Skin: Juliette de Bairacli Levy often described to me how she loved rinsing her face with an elderflower wash. Elderflower's astringent and emollient properties are effective in soothing skin inflammations and healing cuts, wounds, and ulcers. Years ago a naturopathic doctor asked me to make a crème for a child with eczema. I included elderflowers in this crème. The use of a homeopathic remedy and the elderflower crème cleared up the child's eczema. Avena still makes this crème as it is so effective for healing various types of skin eruptions, itchy skin, and rashes.

INDICATIONS FOR BERRIES

Circulatory: Elderberries are high in proanthocyanidins (PCOs), which are known to strengthen arteries, veins (including varicose veins and spider veins), and capillaries. I combine elderberries with lycium berries (*Lycium chinensis*) to help stabilize the small capillaries in the eyes and assist people with poor night vision and macular degeneration. Elderberries along with blueberries, black raspberries, and black currants contain different amounts of PCOs. Foods high in PCOs are known to inhibit reactive oxygen species (ROS), preventing free radical damage and oxidative diseases such as cancer.

Immunity: The berries can be used alone or in combination with elderflowers for preventing and resolving viral flu infections. Years ago I read research from Israel documenting how elderberry juice stimulates the immune system and inhibits different influenza viruses. Madeleine Mumcuoglu, PhD, from the Hadassah-Hebrew University Medical Center in Ein Kerem, Israel, discovered that the constituents contained in elderberries "disarm" a specific enzyme known as the neuraminidase enzyme within 24–48 hours, stopping the replication of the virus (Phillips, 2011, p. 382). Clinical trials with elderberry products have shown that elderberry both prevents and resolves cold and flu symptoms. This has certainly been my experience. Before I planted elder shrubs at Avena I used to travel north to a friend's farm to collect the flowers and berries. My friend worked as a nurse for many years in the local hospital. She took an interest in elderberries, made a gallon of elderberry medicine, and generously passed it out to her co-workers to keep them well during flu season. It worked. She made elderberry medicine for the nursing staff for years. May nurses like Mary-Anne continue to serve their communities and educate their clients about the benefits of medicinal foods like elderberries.

PREPARATION AND DOSAGE

Infusion: A handful of fresh flowers can be placed in a glass or stainless pot. Cover with water and slowly warm. Turn off heat and infuse, covered, for 20 minutes. Place 2 teaspoons of dried flowers into a pot and cover with 8 ounces of hot water. Let infuse for 20–30 minutes. Drink 2–3 cups per day.

Glyerite: (1:4) Fresh or dried flowers. Use ¼–1 teaspoon, 3–5 times per day.

Tincture: (1:4) Fresh or dried flowers or berries. Use ¼–1 teaspoon, 3–5 times per day.

Infused oil: Infuse fresh or dried elderflowers in an organic carrier oil of your choice for two weeks at 100°F. Pour off the oil through unbleached cheesecloth and store in a glass jar in a cool, dark cupboard. This oil can also be made into a salve or crème.

Safety considerations: Avoid eating large amounts of the fresh berries, as the seeds contain small amounts of cyanogenic glycosides, which can cause nausea and diarrhea. Freshly tinctured, cooked, or dried berries are not toxic.

Long-term safety: safe

Elderberry Rosehip Syrup

10 cups fresh or frozen elderberries
10 cups fresh, sliced rosehips
5, 2-inch cinnamon sticks
5 inches finely chopped fresh ginger root
20 cups water
4 cups raw honey
10 tablespoons of organic lemon juice
2 ½ cups high-quality brandy

Combine elderberries, rosehips, cinnamon, ginger, and water. Cook down to 10 cups (half the volume). Strain and cool. Add honey, lemon juice, and brandy. Store in glass jars in a cool, dark cupboard. Shake before using as it is thick. Take a few teaspoons as needed to prevent or treat a cold or flu.

—DEBORAH WIGGS, *East Blue Hill, Maine*

HAWTHORN

Latin names: *Crataegus laevigata, C. monogyna, C. oxycanthoides, C. phaenopyrum*

Common names: Mayflower, May blossom, whitethorn

Family: Rosaceae

Place of origin: Many different species are native to Asia, North America, and Europe.

Parts used: berry, flower, and leaf

Energy: slightly warm

Flavor of berries: sour

Flavor of flowers: sweet, slightly bitter

Cultivating: Years ago while visiting friends in England and Ireland, I observed hawthorn and elder shrubs growing in hedgerows along stone walls. This inspired me to plant a hedgerow of hawthorn trees around the first garden I planted at Avena's farm. In 1997 I purchased 50 young Washington Hawthorn trees from the FEDCO tree catalog and planted them 10 feet apart, creating a natural living fence and deer barrier. These trees are now 15 feet tall and offer up dozens of pounds of blossoms and berries each year. I love standing beneath them when they're blooming, listening to the thousands of honey bees busy at work collecting nectar and pollen.

(The FEDCO tree catalog includes valuable advice on planting and tending fruit trees, as does Michael Phillips's fabulous book *The Holistic Orchard*.)

Collecting: I use a locally made tripod-shaped wooden apple ladder for collecting hawthorn flowers. (This ladder is hand-crafted by Peter Baldwin in Brooks, Maine. He sells his high-quality wooden apple ladders throughout the Northeast.) With a small berry basket strung around my neck and Joyce Chen scissors in my pocket, I climb the apple ladder and carefully cut the newly opened hawthorn flowers into my basket. These flowers are processed immediately into fresh tinctures and carefully laid out on screens to dry.

In October the trees are covered with small red berries. I use a long apple pruning pole for cutting the various branches hanging heavy with the berries. Once the branches are on the ground, we use Felco #310 shears to cut the berries off. Avena makes several gallons of tincture from the fresh berries. I also dry berries for wintertime teas.

Actions: anti-inflammatory, astringent, cardiac tonic, coronary and peripheral vasodilator, circulatory tonic, diuretic, hypotensive, relaxant

INDICATIONS

Cardiovascular: Hawthorn flowers, leaves, and berries are frequently used by medical doctors, especially in Europe, for the benefit it offers the cardiovascular system. Hawthorn improves heart muscle function, increases diameter of coronary artery, stabilizes collagen, lowers mildly elevated blood pressure, reduces inflamed connective tissues, and supports post-surgical recovery. In Ayurvedic medicine hawthorn is used to lower cholesterol and strengthen the cardiovascular system. Hawthorn is used by many practitioners to reduce hardening of the arteries and support a person experiencing an irregular heart beat, heart palpitations, or atrial fibrillation. Hawthorn is used to lower triglycerides and LDL and VLDL cholesterol levels and to support recovery from myocardial infarctions. Hawthorn improves overall circulation and when used regularly can be helpful for people with cold hands and feet, including Raynaud's. Hawthorn is known to increase the uptake of oxygen to the heart and mildly dilate both coronary and peripheral vessels. This action brings more blood to the heart and improves circulation.

Digestion: Flowers, leaves, and berries relax and stimulate the digestive system. Anne McIntyre says hawthorn clears toxins from the gut and helps regulate metabolism.

Nervous system: Hawthorn helps calm and settle a person who feels anxious or agitated. I have seen many people who were unable to sleep because of grief, sadness, and anxiety receive relief when using a hawthorn tincture over several months. I often combine hawthorn with heartsease pansy, motherwort, and lemon balm for people who are experiencing grief, loss, or depression. Many Western herbalists combine hawthorn with lemon balm for children and adults with ADD and ADHD. Hawthorn in combination with other herbs can be helpful for people who suffer from posttraumatic stress disorder.

Tissues: Hawthorn stabilizes collagen and arteries. Its antioxidant properties ease inflammatory connective tissue disorders and support the healing process following surgery. Hawthorn can be combined with gotu kola, sarsaparilla, and turmeric for healing inflammatory connective tissue disorders such as lupus, scleroderma, and rheumatoid arthritis (Kuhn and Winston, 2001, p. 256).

PREPARATION AND DOSAGE

Flower infusion: Infuse 1 tablespoon of dried flowers in 8 ounces of hot water for 15–30 minutes. Place a handful of fresh flowers into a glass or stainless pot, cover with cool water, and slowly heat. Turn off heat before water boils. Infuse for 15–30 minutes.

Berry infusion: Place 1–2 tablespoons of fresh or dried berries into a pot and simmer in 8 ounces of water for 5–10 minutes. Turn off the heat and steep for 30–60 minutes.

Powder: Mix ½–1 teaspoon of powdered berries in ½ cup of warm water, taken once or twice a day.

Tincture: (1:5) Fresh or dried flowers. Fresh or dried berries. Take ¼–1 teaspoon, 3–5 times per day.

Hawthorn berry solid extract: A concentrated, non-alcohol extract of hawthorn berries. Take ¼–½ teaspoon directly under the tongue or mixed in ¼ cup of warm water, twice daily.

Safety considerations: Use hawthorn under the guidance of a knowledgeable health care provider if taking pharmaceutical cardiac medications. Hawthorn may potentiate actions of heart medications such as beta blockers.

Long-term safety: safe

Sweet Cicely

Sweet cicely (*Myrrhis odorata*) has begun to naturalize under many of our hawthorn trees, creating a living mulch and providing nectar to the honey bees as it blooms three to four weeks earlier than the hawthorn trees. Sweet cicely leaves, collected fresh, make a lovely licorice-tasting tea. It's a hardy perennial that reseeds abundantly, grows in full sun and partial shade, and appreciates compost.

OATS (green milky)

Latin name: *Avena sativa*

Common name: oats

Family: Poaceae, formerly Graminaceae

Place of origin: found worldwide in fields and farmlands of both temperate and cooler temperature zones

Part used: green, milky oat seed

Energy: warm to neutral

Flavor: sweet

Cultivating: Organic oat seeds are available through many seed companies. Be sure the seed is certified organic. Nonorganic seed is often treated with a fungicide. Oats are a valuable cover crop and medicinal herb and can be planted between early May and mid-June in zone 5. Each year we sow a few beds of oats in different parts of the garden depending on which garden beds are ready to be taken out of production for a year. Prepare a seed bed and scatter the oat seeds. I use a level-head rake to gently scratch the seeds into the soil; I look for a thick, green carpet of oat sprouts to appear within 7–10 days. Watch out for critters eating oat seeds—wild turkeys in particular.

Collecting: The weather always plays a role when waiting for an herb to be ready for collecting. We hand gather the green milky oat seeds, usually two months after sowing. I start checking the oat seeds for their milk content one to two weeks before we gather them. As soon as the seeds are green and plump and a white milklike substance bursts forth from squeezed seeds, they're ready to be collected. The oat seeds gathered from Avena's garden are immediately made into fresh oat tinctures and glycerites. The oat seeds we dry for Avena's teas are gathered from a nearby biodynamic vegetable farm, Hope's Edge, where they plant ¼–½ acre of certified organic oats as part of their farm's rotation plan. Avena staff hand harvest 300 pounds of fresh oat seeds, collecting them into large baskets that are then transported back to Avena's drying room and spread out on screens to dry. It takes about 10 days, drying at 80°F, before the seeds are completely dry. Consistent heat between 70°F and 80°F is essential when drying oat

seeds. Drying at too hot of a temperature will dry the outer coat quickly and yet not completely dry the inner seed, which can lead to spoilage once they're bagged up and stored. Careful attention is required for those who tend herbs in drying rooms.

Actions: antidepressant, nutritive, restorative tonic to the nervous system

INDICATIONS

Gynecology: Fresh oat seed tincture is beneficial for women to use orally on a daily basis over several months for moistening and lubricating dry, irritated, or inflamed vaginal tissue. Oat seed tincture combined with white pond lily tubers, lady's mantle, and shatavari root is beneficial to women in their menopausal and postmenopausal years for lubricating and moistening dry vaginal tissue.

Immunity: Daily use of fresh oat seed tincture or glycerite supports the body's healing when outbreaks of chicken pox, herpes, and shingles occur. Use during both the acute and recovery stages.

Nerves: Green milky oat seed tincture or glycerite restores strength and vitality to the nervous system. It's my favorite tonic herb for the nervous system, which is why I named my business Avena Botanicals. When taken over several weeks or months, oats' ability to ease anxiety, nervous debility, fatigue, depression, insomnia, premenstrual and menopausal stress, and physical and mental exhaustion (along with lifestyle considerations) is remarkable. Oats are a key herb to include in formulas for people (and their family members and caregivers) who are living with chronic degenerative conditions. Oats are useful during times of high stress, especially in working environments when people feel emotionally drained and easily reactive. Many herbalists recommend using oats as part of a protocol for helping reduce the cravings and agitation people experience when withdrawing from nicotine, caffeine, alcohol, drugs, or addictive behaviors.

Skin: Oatmeal poultices and baths soothe dry and itchy skin irritations and inflammatory skin conditions.

PREPARATION AND DOSAGE

Infusion: Fresh oat seed tea can be made by first collecting a large handful of the green milky seeds. Place the seeds in a glass or stainless pot, cover with water, and slowly bring to a simmer. Turn off the heat and let infuse for 15–60 minutes. Fresh milky oat seed tea is sweet and feels deeply nourishing to the whole body. To make an infusion from dried oat seeds, simmer 1–2 tablespoons of the seeds in 8 ounces of water for 5 minutes, covered. Let steep for 10–30 minutes.

My mind becomes quieter when I imagine the gentle sound of the oat seeds blowing in the wind. This image reminds me to stop and sit quietly while drinking a cup of herb tea. The nervous system and Spirit are nourished when we really stop and listen to the sounds in nature and breathe in the beauty of Earth. Quietly sipping a warm cup of tea when sitting in the garden or looking out the kitchen window can be helpful as a daily practice for relaxing the body, mind, and spirit.

Glycerite: (1:2) Fresh, green milky oat seeds. Take ¼–1 teaspoon, 3–5 times per day.

Tincture: (1:2) Fresh, green milky oat seeds. Take ¼–1 teaspoon, 3–5 times per day.

Safety considerations: Recent studies show that the majority of people with a gluten intolerance can eat oats that are processed in a gluten-free environment.

Long-term safety: Safe; long history of use as food.

SCHISANDRA

Latin name: *Schisandra chinensis*

Common names: magnolia vine, Chinese magnolia vine

Family: Magnoliaceae

Place of origin: China

Part used: fruit and seeds

Energy: warm, dry

Flavor: Sweet, sour, salty, bitter, and pungent. The Chinese name, *wu wei zi,* means "five flavors fruit."

Cultivating: All but one of the 25 known *Schisandra* species are native to the forests of northern China, the Russian Far East, Korea, and Japan. *Schisandra coccinea,* commonly referred to as southern magnolia vine, is a rare species found growing in undisturbed stream beds in North Carolina, Tennessee, Georgia, Florida, Arkansas, and Louisiana (Foster and Chongxi, pp. 146–147). Schisandra vines prefer well-drained and deeply cultivated sandy soil, plenty of moisture, and rich compost. Maine and northern New England are good climates for establishing these perennial vines, which appear to be free of diseases and pests. But do be aware of chipmunks eating the fresh berries in fall! They leave little piles of half-eaten berries on the stone wall adjacent to one of our large cedar arbors. Like most vines, schisandra needs a sturdy arbor or fence on which to grow. It usually begins bearing fruit after the fourth year. In Avena's gardens the schisandra vines growing in full sun began producing fruit three years earlier than the vines growing in partial shade. We are currently attempting to establish 18 new vines on a sunny hillside location where they will be trained to grow like grapes along our deer fence.

Schisandra chinensis is the species most commonly listed in medicinal herb books. The small, lightly fragrant, white flowers grow in clusters, and in our garden they attract thousands of honey bees. The female and male flowers grow on separate plants, and both are needed to produce fruits. These flowers give way to beautiful red bunches of berries in the fall. The berries are highly valued for their medicinal properties.

Schisandra is propagated by seed, stem cuttings, root suckers, or layering. Well-established vines send out numerous runners at their base. These runners can be cut in late spring and moved to a newly prepared bed that contains plenty of compost and mulch. Water the transplants every few days as they establish themselves and then at least once a week during their first growing season. (You can mail-order female and male schisandra vines from One Green World, www.onegreenworld.com. Peg Schafer's book *The Chinese Medicinal Herb Farm,* published in 2011, is an excellent reference for people who wish to grow and harvest Chinese medicinal herbs.)

I am on a mission to encourage as many people as possible living in northern climates, especially zones 4–5, to organically cultivate schisandra vines. It has the potential to be a valuable crop for gardeners and farmers in Canada and the United States to grow and sell to herbalists. It's also the perfect perennial fruit for permaculture practitioners to grow for their family and community.

Collecting: Eating a freshly picked berry is fabulous. There is just no comparison between sucking on a fresh berry versus a dried one—though both are highly medicinal. We collect the fresh berries in early fall, once they're ripe, and we tincture them immediately. A tincture made from fresh berries is a vibrant red color with a hint of magenta. The berries can be processed into a fresh juice or dried.

Actions: adaptogen, anti-asthmatic, anti-inflammatory, astringent, hepatoprotective, immune tonic, memory tonic, nervine

Adaptogen: Schisandra is used by Western herbalists as an overall tonic and restorative herb for numerous deficient and weak conditions in the body. It's effective when combined with other herbs for easing chronic stress, chronic fatigue, insomnia, poor memory, depression, fear, night sweats, and involuntary sweating. The berries help to regulate blood sugar levels and improve overall stamina and endurance. Schisandra can be used for several weeks or months to improve energy levels, reduce tiredness, strengthen the immune system, and rebuild resiliency. An important herb for anyone who has experienced trauma to support their healing process. Herbalist David Winston says that schisandra returns neurotransmitter hormones to normal more quickly after an adrenalin rush. I add a few droppers of Avena's fresh Schisandra berry tincture to my quart water bottle throughout the gardening season as part of my tick prevention protocol, helping to keep my immunity and adrenals healthy and my spirit peaceful. (see pg 199 tips for gardeners during tick season)

INDICATIONS

Cardiovascular: Schisandra has a calming effect on the body and can help reduce stress-induced heart palpitations. It's commonly used in Traditional Chinese Medicine to calm *shen*. The Chinese believe that *shen* (spirit, consciousness) is stored in the heart. Disturbed *shen* can cause symptoms such as agitation, anxiety, palpitations, insomnia, and impaired memory.

Gynecology: I use schisandra in formulas for women experiencing menopausal sweating, stress-induced heart palpitations, insomnia, memory loss, and anxiety. Schisandra's astringent qualities can help a woman feel held and supported when experiencing major life transitions like menopause or the death of a family member or close friend.

Kidneys: Schisandra's astringent nature and ability to nourish the kidneys may be helpful for people who urinate frequently, have early morning diarrhea, or night sweats. If these conditions persist, a visit to an herbalist or Chinese practitioner can help address the root causes of these symptoms.

Liver: Schisandra berries help regenerate hepatocytes and protect the liver when the body has been exposed to toxic chemicals or drugs. Combine the berries with milk thistle seed and turmeric root for preventing or treating liver damage caused by industrial solvents, pharmaceutical medications, recreational drugs, alcohol, and viruses, including hepatitis B (Winston and Kuhn, 2001, p. 297).

Respiratory: Schisandra benefits various respiratory weaknesses. Traditional Chinese herbalists say the fruit stabilizes lung qi. Its astringent and tonifying properties help resolve chronic coughs and allergic asthma that is accompanied with wheezing. I use schisandra in formulas for people who have weak lungs and are prone to respiratory infections. I also recommend people begin using schisandra one to two months ahead of the time of the year they commonly experience a respiratory infection. People with reoccurring infections or chronic tiredness may benefit by consulting with an acupuncturist or Ayurvedic practitioner. These practitioners are trained to use the pulse and tongue as diagnostic tools, allowing them to treat the underlying causes of a person's illness rather than just treating symptoms.

PREPARATION AND DOSAGE

Decoction: 1–2 tablespoons of fresh or dried berries simmered in 8–12 ounces of water for 5–10 minutes. Infuse for 15–30 minutes and drink 1–2 cups per day.

Juice: Fresh berry juice can be made by using a steam juicer. You can freeze or process the juice in canning jars for winter use.

Powder: ½–1 teaspoon of freshly ground powder mixed with a ½ cup of warm water, taken once or twice a day.

Tincture: (1:5) Fresh or dried berries, ¼–½ teaspoon, 3–4 times per day.

Safety considerations: Avoid during pregnancy, as the berries may stimulate uterine contractions. Berries are considered safe for nursing women.

Avoid during acute fevers, flu, bronchitis, or pneumonia or any acute condition with excess heat such as skin rashes. Schisandra is contraindicated for people and animals with epilepsy and must be used carefully for anyone who is on a phenobarbitol or barbital, as the berries may potentiate the action of those drugs (Kuhn and Winston, 2001, p. 298).

Long-term safety: safe

OPPOSITE PAGE: A song sparrow hanging out on a Greek Mullein stalk.

Wu Wei Zi Schizandra:

Wu Wei Zi Schizandra is in the Astringent category in Chinese Herbal Medicine. Astringent herbs hold the fluids and structures in place. Wu Wei Zi Schizandra treats the Lung, Heart and Kidney, and helps us hold onto what is eternally valuable in life. For example, we are able to hold the essence of the people we love close to us regardless of whether or not the person is actually still present because of the Lung. The Lung also receives the breath qi and inspiration from Heaven. Wu Wei Zi astringes the Lung to keep what is valuable (essence and qi) from leaking out in the form of respiratory problems such as cough and wheeze. The Kidney makes an appearance here as it helps "grasp" the qi making it possible to take a nice deep breath without coughing it back before receiving all the qi. Wu Wei Zi also aids the Kidney to stop the leakage of fluids (valuables) from the other end, treating diarrhea, incontinence, and spermatorrhea. The Heart in Chinese Medicine is "Lord and Sovereign" and is the residence of the Spirit (Shen, your Divine nature). If the Lord and Sovereign of your Body/Mind/Spirit is to bring your truest self into Life, it must provide an inner sanctuary from which the Spirit can shine its light. Wu Wei Zi holds the Heart steady so your light can shine.

-Amy Jenner, Acupuncturist and Chinese Herbalist

Everybody needs beauty as well as bread, places to play in and pray in, where nature may heal and give strength to body and soul alike.
— JOHN MUIR

AFTERWORD

*I'm going to leave a heart in the earth so that it may grow and flower
and adore everything green.*

—Rosario Murillo

I recently read the second edition to Vandana Shiva's book *Staying Alive: Women,
Ecology and Development* (first published in 1988). Vandana Shiva, along with
Rachel Carson, Terry Tempest Williams, and Sharon Salzberg, are some of the
women I respect deeply. Vandana describes the connections between globalization,
ecological crises, colonialism, and the oppression of women in stunningly honest ways.
As this book goes to press, food security policies, the battle over labeling genetically
modified food, violence against women, gun control, and health care are in the news in
the United States. All are seriously troubling. I read Vandana Shiva because she offers a
larger perspective, and sustainable solutions, to the multitude of problems of our time.
I meditate and pray for the well-being of every living being because this daily Metta
practice helps my heart remain open in the face of such suffering on our planet.

As a woman, I want to understand the challenges of women farmers worldwide.
Vandana writes:

> Women farmers in the Third World are predominantly small farmers. They provide the
> basis of food security, and they provide food security in partnership with other species.
> This partnership between women and biodiversity has kept the world fed throughout
> history, is doing so in the current moment, and will continue to feed the world in the
> future. It's this partnership that needs to be preserved and promoted to ensure food
> security.

> Agriculture based on diversity, decentralization, and improving small farm productivity
> through ecological methods is a women-centered, nature-friendly agriculture. In this
> agriculture, knowledge is shared—other species and plants are kin, not "property"—
> and sustainability is based on the renewal of the earth's fertility and the renewal and
> regeneration of biodiversity and species richness on farms. There is no place for
> monocultures of genetically engineered crops…When this technology was introduced

into the Third World farming systems it led to an increased use of agrochemicals, thus increasing environmental problems. It also destroyed the biodiversity that is the sustenance and livelihood base of rural women: what are weeds for Monsanto are food, fodder, and medicine for Third World women.

Grow food and medicinal herbs in "nature-friendly" ways. Call upon the plants to keep yourself and your family well. Share them with your neighbors. Learn to identify edible and medicinal weeds and the trees, birds, and pollinators you live amongst. Share this knowledge with your community. Know your local farmer and herbalist and support them. Let kindness and compassion grow within your own heart and inspire your connection with all beings everywhere.

For more information on how to prepare and use plant medicines, and for yearly listing of classes and herb walks offered by Deb, refer to: **www.debsoule.com**

ACKNOWLEDGMENTS

Many hugs to my family, friends, teachers, students, and co-workers for your love, care, and encouragement. Special thanks to Louise Berliner, Richo Cech, Squidge Davis, Jean-David Derreumaux, Heather Ensworth, Connie Evans, Elizabeth Garber, Kate Gilday, Jennifer Greene, Tom Griffin, Missy Hatch, Amy Jenner, Jillian Kelsey-Rose, Liz Lauer, Kathleen Maier, Selkie O'mira, Chris Shrum, Sue Szwed, Priscilla Skerry, Maia Toll, Michelle Walker, and Julia Yelton for the various ways you supported me in this four-year writing journey. Special thanks to Rocio Alarcon, Paul Bergner, Mary Bove, Rosemary Gladstar, Anne McIntyre, Karyn Sanders, David Winston, and Matthew Wood for your herbal teachings. Deep appreciation to Russell Comstock, CR Lawn, Katey Branch and the late Alan Day for believing and supporting my vision of Avena. A deep bow to the professional photographers G. Michael Brown, Mary Crowley, Susie Cushner, Lynn Karlin, and Stephen Orr for the beautiful photographs you took in Avena Botanicals gardens. Thank you Malcolm Brooks for encouraging me to give the book its title. Many thanks to David Allen, Sally Jaskold, Ed Miller, and Tim Seymour for your professional editing, design, and production assistance. Deep gratitude to Tom Griffin for being my biodynamic buddy, favorite vegetable farmer, and partner.

A tender thank you to the late Russell Libby, executive director of the Maine Organic Farmers and Gardeners Association (MOFGA), for valuing herbalists and our contributions to the farming community. Thank you Robert Karp and Thea Marie-Carlson, co-directors of the national Biodynamic Farming and Gardening Association (BDA), for all of your work and for your support and interest in the well-being of Avena Botanicals. I am grateful to live in Maine, a place where organic and biodynamic farming practices are respected. The BDA is the oldest organic farming organization in the United States and MOFGA is the second oldest. May the work of these organizations, and similar ones, and the work of herbalists and farmers everywhere benefit our vulnerable and beautiful planet.

GLOSSARY

Acute disease: Any disease with sudden onset, severe symptoms, and a brief duration.

Adaptogen: An herb that helps the body adapt instead of react to stress, normalizes stress hormone levels, and restores balance and vitality.

Alterative: An herb that alters the body's metabolic processes. Many alterative herbs help the body to eliminate waste via the liver, lungs, kidneys, or skin.

Ama: An Ayurvedic term that refers to undigested food or life experiences that lead to disease-forming toxins in the body.

Amenorrhea: Absence of menstruation.

Amphoteric: An herb that normalizes organ or tissue functions.

Analgesic: A pain-relieving substance.

Antibacterial: Substances that eliminate or suppress the growth or reproduction of bacteria.

Anti-inflammatory: Substances that relieve inflammation.

Antimicrobials: Herbs that help the body eliminate pathogenic microorganisms such as bacteria, fungi, and viruses through different mechanisms.

Antioxidants: Substances that inhibit oxidation and trap free radicals. This action prevents damage to cells.

Antipyretic: Reduces fevers.

Antiseptic: Destroys or inhibits harmful bacteria.

Antispasmodics: Herbs that relieve muscle spasms and tension in the musculoskeletal system and the smooth muscle of the hollow organs (uterus, bladder, stomach, intestine, gallbladder).

Antiviral: Herbalist Paul Bergner writes in an article titled "Antiviral Botanicals in Herbal Medicine": "The organism possesses multiple mechanisms of host defense to maintain ecological balance in the presence of viruses. Except for the occasional use of herbal medicines as topical agents in viral infections of the skin, plant medicines probably do not exhibit direct virucidal properties within the organism. Many plants have been used traditionally to enhance host resistance to viral infection, and recent advances in immunology have uncovered possible novel mechanisms for their action." http://medherb.com/eletter/Antiviral-Bergner.pdf

Anxiolytic: Reduces anxiety.

Aromatic: An herb with a pleasant or stimulating scent.

Arteriosclerosis: A thickening and calcification of the walls of the arteries causing a loss of elasticity.

Astringent: Contracts, tightens, or tones tissue. Reduces body secretions.

Atherosclerosis: A common arterial disease in which raised areas of degeneration and cholesterol deposits (plaques) form on the inner surfaces of the arteries.

Bitter: Bitter-tasting herbs that have a stimulating effect on the digestive system.

Brain tonics: Herbs that enhance concentration and cognition and improve memory.

Bronchodilator: Improves breathing by relaxing the smooth muscles in the lungs and widening the air passages.

Cardiac tonics: Herbs that strengthen, tone, and improve the function of the heart and blood vessels.

Carminatives: Herbs that help expel gas and relieve pain in the gastro-intestinal tract.

Cervicitis: Inflammation of the cervix caused by infections such as candida, staph., strep., and trichomonas.

Cholagogue: Stimulates the flow of bile from the liver.

Cholecystitis: An acute or chronic inflammation of the gallbladder.

Choleretic: Stimulates the flow of bile into the duodenum.

Chronic disease: A disease with a gradual onset and long-term symptoms.

Circulatory stimulant: An herb that stimulates the flow of blood to various tissues.

Cirrhosis: A degenerative liver disease caused by excess fat and fibrous tissue.

Coronary vasodilator: A substance that dilates the coronary blood vessels.

Decoction: A type of herbal tea prepared by simmering or boiling the herb in water.

Decongestants: Herbs that reduce congestion or swelling in the sinuses, nasal passages, and lungs.

Demulcent: Soothes irritated or inflamed tissues or mucous membranes.

Diaphoretic: Promotes perspiration.

Diuretic: Increases urination.

Dysbiosis: Disturbed gut flora.

Dysmenorrhea: Painful menses, often with uterine cramping.

Emmenagogue: Promotes menstrual flow.

Emollients: Herbs that are used topically to soothe and soften irritated skin and mucous membranes.

Endometriosis: Abnormal growth of the endometrial tissue outside of the uterus.

Expectorants: Herbs that encourage coughing and loosen and clear phlegm from the lungs.

Fibrocystic: A cystic lesion found within fibrous connective tissue.

Fomentation: A warm herbal compress made by dipping a cloth in a hot herbal tea.

Free radical: A highly reactive molecule that joins with an unpaired electron and causes damage to tissues.

Galactogogues: Herbs that promote the flow of milk.

Goitre: A swelling in the neck caused by an enlarged thyroid gland.

Hepatics: Herbs that stimulate or support liver function.

Hepatoprotectives: Herbs that protect and prevent damage to the liver.

Hypertension: High blood pressure.

Hypoglycemic: A substance that reduces blood glucose levels in the blood.

Hypotensive: A substance that lowers blood pressure.

IBS: Irritable bowel syndrome. Symptoms include bloating, gas, constipation, cramping, and diarrhea.

Immunosuppressive: A substance that suppresses the function of the immune system.

Immunomodulator: A substance that regulates immune functions by either augmenting or diminishing the immune system's response.

Immunostimulant: Stimulates the function of the immune system.

Infusion: An herbal tea made by steeping or infusing the herb in water.

Kidney tonic: Improves the function of the kidneys and renal system.

Laxative: Stimulates bowel peristalsis.

Lipomas: Fatty tissue that forms a benign tumor.

Lung tonics: Herbs known to strengthen and tone the function of the lungs.

Lupus: Systemic lupus erythematosus. An autoimmune condition.

Lyme disease: An infection caused by the spirochete *Borrelia burgdorferi* that enters the body via a tick bite.

Lymph: Clear fluid that flows through the lymph vessels. It nourishes tissue cells and returns waste material to the bloodstream.

Lymphadenitis: Inflammation of the lymph nodes.

Lymphatic: An herb that stimulates the flow of the lymphatic system.

Mastitis: A breast infection often caused by the bacteria *Staphylococcus aureus*.

Menorrhagia: Heavy menstrual bleeding that is considered abnormal.

Nervine: A category of herbs that relax the nervous system and reduce stress and tension.

Nutritives: Herbs and foods that nourish the body and often contain high amounts of bioavailable minerals and vitamins.

PMS: Premenstrual stress.

Prana: The universal life force, or vital energy, which exists in all things.

Proanthocyanidins: aka oligomeric proanthocyanidins (OPC), is in the class of antioxidants known as flavonoids.

Qi (chi) tonics: Tonic herbs are those that strengthen or nourish a particular area or function of the body that has become weakened. Qi tonics treat patterns where a deficiency of qi has resulted in problems arising out of an organ not having the energy it needs to carry out its functions. Qi deficiency primarily affects spleen and lung functions, as those are the organs charged with extracting qi from food and air, transforming it into a usable form, and distributing it throughout the body.

Rejuvenatives: Herbs that restore vitality and overall well-being.

Relaxants: Herbs that relax the nervous system.

Restoratives: Herbs that restore health and homeostasis to the body.

Spleen qi: The primary function of the spleen is that it extracts qi from food and transforms it into a substance that can be used by the body to make your qi. It then distributes that qi throughout the body. Having adequate spleen qi results in the appropriate appetite and digestion, energy, strength, and clear thinking to support a rich life. This includes extracting qi from life experiences and intellectual pursuits and making them part of you.

Stimulant: A substance that stimulates the central nervous system and increases metabolism.

Tinea: Fungal skin infections such as athlete's foot, ringworm, and jock itch.

Tonic: Tonic herbs restore health and normal function to organs and tissues. They're most effective when used over a period of 1–3 months or longer if needed. Tonic herbs are best avoided during an acute illness and resumed once the illness is resolved.

Trophorestorative: An herb or food that acts as a nutritive restorative for the body, usually with an affinity for a specific organ or organ system. Trophorestoratives resolve and eliminate deficient conditions and weaknesses by nourishing and rebuilding the health and vitality of the organ or organ system.

Uterine tonic: An herb that tones the uterine muscles and improves overall uterine health.

Vaginitis: Inflammation of the vaginal tissue.

Vasodilator: A substance that widens the blood vessels and lowers blood pressure.

Vulneraries: Herbs that promote the healing of wounds.

SEED AND PLANT SOURCES

Abundant Life Seeds
PO Box 157
Saginaw, OR 97472-0157
www.abundantlifeseeds.com

Fedco Seeds
PO Box 520
Waterville, ME 04903-0520
www.fedcoseeds.com
Fedco has four divisions: Seeds,
Moose Tubers and Organic
Growers Supply, Trees, and Bulbs.

High Mowing Organic Seeds
76 Quarry Rd
Wolcott, VT 05680
www.highmowingseeds.com

Johnny's Selected Seeds
955 Benton Ave.
Winslow, ME 04901-2601
www.Johnnyseeds.com

One Green World
6469 SE 134th Ave.
Portland, OR 97236
www.onegreenworld.com
They ship unusual varieties of fruit
and nut trees and a few medicinals.

Prairie Moon Nursery
32115 Prairie Lane
Winona, MN 55987
www.prairiemoon.com

Seed Savers Exchange
Rt. 3, Box 239
3094 North Winn Rd.
Decorah, IA 52101
www.seedsavers.org

St. Lawrence Nurseries
325 State Hwy. 345
Potsdam, NY 13676
www.sln.potsdam.ny.us
They ship northern climate fruit
and nut trees.

Strictly Medicinals
PO Box 69
Williams, OR 97544-0069
www.horizonherbs.com
They mail order medicinal plants
and trees.

**Turtle Tree Biodynamic Seed
Initiative**
10 White Birch Rd
Copake, NY 12516
www.turtletreeseed.org

Zack Woods Herb Farm
278 Mead Road
Hyde Park, VT 05655
www.zackwoodsherbs.com
They mail order certified organic
plants and dried herbs. Melanie and
Jeff Carpenter's book The Organic
Medicinal Herb Farm (published
in 2015) is an excellent resource for
small and large herb growers.

NATIVE AMERICAN BASKETS AND RESOURCES
Recommended by Jennifer Neptune

Books:

McBride, Bunny, and Harold E. L. Prins. *Indians in Eden*. Camden, ME: Downeast Books, 2009.

McBride, Bunny. *Women of the Dawn*. Lincoln, NE: Bison Books, 2001.

Neptune, Jennifer. "Wabanaki Traditional Arts: From Old Roots to New Life," from *North by Northeast* by Kathleen Mundell. Gardiner, ME: Tilbury House, 2008.

The Wabanakis of Maine and the Maritimes. An excellent resource published by the American Friends Service Committee.

Websites:

Maine Indian Basketmakers Alliance
www.maineindianbaskets.org

Maine-Wabanaki REACH is a collaborative program that works towards truth, healing and change in Wabanaki and Maine communities. In Wabanaki communities the focus is on health, wellness and self-determination. In Maine communities, the focus is on education about our collective histories and moving forward in decolonizing ourselves and our organizations and systems. www.mainewabanakireach.org

Penobscot Nation Cultural & Historic Preservation Department
www.penobscotculture.org

Museums:

The Abbe Museum, Bar Harbor, Maine: www.abbemuseum.org

The Heard Museum, Phoenix, AZ: www.heard.org

The Hudson Museum, Orono, Maine: www.umaine.edu/hudsonmuseum

The Maine State Museum, Augusta, Maine: www.mainestatemuseum.org

Indian Township Museum, Indian Township, Maine 207-796-5533

National Museum of the American Indian. Washington DC. http://nmai.si.edu/visit/washington/

Penobscot Nation Museum, Indian Island, Maine: www.penobscotnation.org/museum 207-827-4153238

Wabanaki Museum & Resource Center: www.wabanaki.com/museum.htm 207-853-2600 ext 227

Native American and Diversity Resources: *Recommended by Karyn Sanders*

A Different Mirror: A History of Multicultural America by Ronald Takaki, 1993.

Gathering of Spirit: A Collection of North American Indian Women ed. Beth Brant, 1984.

Hope Dies Last: Keeping the Faith in Troubled Times by Studs Terkel, 2003.

My Sisters' Voices: Teenage Girls of Color Speak Out by Iris Jacob, 2002.

Out of the Class Closet: Lesbians Speak ed. Julia Penelope, 1994.

Prison Writings: My life Is My Sundance by Leonard Peltier, ed. Harvey Arden, 2000.

Reinventing the Enemy's Language: Contemporary Native Women's Writings of North America eds. Joy Harjo and Gloria Bird, 1997.

Sister Outsider: Essays and Speeches by Audre Lorde, 1984.

The State of Native America: Genocide, Colonization and Resistance, ed. M. Annette Jaimes, 1999.

Yearning: Race, Gender and Cultural Politics by bell hooks, 1990.

Yellow: Race in America Beyond Black and White by Frank H. Wu, 2002.

Other recommended resources:

An excellent documentary film titled *Girl Rising* is about the challenges girls face worldwide to receive an education. http://girlrising.com

Kristof, Nicholas D., and Sheryl WuDunn. *Half the Sky: Turning Oppression into Opportunity for Women Worldwide*. New York: Vintage Books. 2010.

dRworks is a group of trainers, educators and organizers working to build strong progressive anti- racist organizations and institutions. **dRworks** can be reached at www.dismatnlingracism. org. Recommended by long time herbalist and activist Kahadish.

Ensler, Eve. *In the Body of the World*. New York, NY: Metropolitan Books. 2013. Eve is also the author of *Vagina Monologues* and the founder of V-day, the global movement to end violence against women and girls called "One Billion Rising." www.vday.org

Sustainable Harvest International A Maine based non-profit working directly with farmers in Central America to plant hope, restore forests, and nourish community.

www.sustainableharvest.org

Biodynamic Resources:

Biodynamic Farming and Gardening Association: www.biodynamics.com

Josephine Porter Institute: www.jpibiodynamics.org

Steiner Books: www.steinerbooks.org

Documentary Films:

An excellent documentary film titled *Girl Rising* is about the challenges girls face worldwide to receive an education. http://girlrising.com

First Light. An 11 minute (is being expanded) documentary showing the past and present challenges many Native American families in Maine and Canada have faced and the powerful Truth and Reconciliation work that recently occurred in Maine to address and support a healing process. www.upstander.org

Tomorrow by French film makers Cyril Dion and Mélanie Laurent. 2015. https://www.demainlefilm.com/en/film

Herbal Films:

Juliette of the Herbs: www.julietteoftheherbs.com

Numen: The Healing Power of Plants: www.numenfilm.com

PERMISSIONS

Grateful acknowledgment is given to the following publishers and authors for their permission to use specific quotes and poems.

Ackerman, Diane. *Cultivating Delight: A Natural History of My Garden.* New York, NY: HarperCollins Publishers, Inc., 2001. Reprinted by permission of the publisher.

CHAPTER I

Pages

7 McGreevy, Joyce. *Gardening By Heart.* San Francisco, CA: Sierra Club Books. 2000. Reprinted by permission of the publisher.

8 Steiner, Rudolf. Translated by George Adams. (first published in English in 1958) *Agriculture Course: The Birth of the Biodynamic Method.* Forest Row, UK: Rudolf Steiner Press. Reprinted by permission of the publisher.

11 Ibid.

11 Johnson, Wendy. *Gardening at the Dragon's Gate.* New York, NY: Bantam Books. 2008. Reprinted by permission of the publisher.

17 Don, Montagu. *The Sensuous Garden.* Simon & Schuster Editions. 1997. Reprinted by permission of the publisher.

21-23 Neptune, Jennifer. "Wabanaki Traditional Arts: From Old Roots to New Life," from *North by Northeast* by Kathleen Mundell. Gardiner, Maine: Tilbury House, Publishers. 2008. Reprinted by permission of the publisher.

25 Machado, Antonio. *Times Alone: Selected Poems of Antonio Machado,* translated by Robert Bly. Copyright 1983. Reprinted by permission of Wesleyan University Press.

25 National Academy of Science. *Status of Pollinators in North America.* Washington, DC. 2007. Reprinted by permission of the publisher.

26 Tara Brach, *Radical Acceptance.* New York: Bantam Books, 2003. Reprinted by permission from the publisher.

26 Buchmann, Stephen L., and Nabhan, Gary Paul. *The Forgotten Pollinators.* Washington, DC: Island Press. 1996. Reprinted by permission of the publisher.

26-27 Nabhan, Gary Paul. *Cross-Pollinations: The Marriage of Science and Poetry.* Minneapolis, MN: Milkweed Editions. 2004. Reprinted by permission of the publisher.

33   Schaefer, Carol. *Grandmothers Counsel the World.* Boston, MA: Trumpeter Books. 2006. Reprinted by permission of the publisher.

CHAPTER II

39   Quote is from The Biodynamic Farming and Gardening Associations quarterly journal.

43 Shiva, Vandana. *Soil Not Oil: Environmental Justice in an Age of Climate Crisis.* Brooklyn, NY and Boston, MA: South End Press. 2008.

45 Steiner, Rudolf. *What Is Biodynamics? A Way to Heal and Revitalize the Earth.* Great Barrington, MA: SteinerBooks. 2005. Reprinted by permission of the publisher.

51   Smith, Richard Thornton. *Cosmos, Earth and Nutrition.* East Sussex, UK: Sophia Books. 2009. Reprinted by permission of the publisher.

53   Spock, Marjorie. *Fairy Worlds and Workers.* Hudson, NY: Anthroposophic Press. 1980. Reprinted by permission of the publisher.

54   Steiner, Rudolf. *Harmony of the Creative Word.* E. Sussex, UK: Rudolf Steiner Press. 2001. Reprinted by permission of the publisher.

57   Wildfeuer, Sherry. *Stella Natura.* Kimberton Hills, PA: Camphill Village Kimberton Hills. 2010. Reprinted by permission of the author.

59   Smith, Richard Thornton. *Cosmos, Earth and Nutrition.* East Sussex, UK: Sophia Books. 2009. Reprinted by permission of the publisher.

61   Ibid.

62   Wright, Hillary. *Biodynamic Gardening: For Health and Taste.* Edinburgh, Scotland: Floris Books. 2007. Reprinted by permission of the publisher.

67   Steiner, Rudolf. Translated by George Adams. First published in English in 1958. *Agriculture Course: The Birth of the Biodynamic Method.* E. Sussex, UK: Rudolf Steiner Press. Reprinted by permission of the publisher.

68   Wright, Hillary. *Biodynamic Gardening: For Health and Taste.* Edinburgh, Scotland: Floris Books. 2007. Reprinted by permission of the publisher.

69   Steiner, Rudolf. Translated by George Adams. First published in English in 1958. *Agriculture Course: The Birth of the Biodynamic Method.* E. Sussex, UK: Rudolf Steiner Press. Reprinted by permission of the publisher.

70   Ibid.

71-72   Ibid.

72   Ibid.

75 Steiner, Rudolf. *What Is Biodynamics? A Way to Heal and Revitalize the Earth.* Great Barrington, MA: SteinerBooks. 2005. Reprinted by permission of the publisher.

76   Wood, Matthew. *The Book Of Herbal Wisdom.* Berkeley, CA: North Atlantic Books. 1997. Reprinted by permission of the publisher.

79   Dawson, Adele G.. *Herbs Partners in Life*. Rochester, VT: Healing Arts Press. 2000. Reprinted by permission of the publisher.

82   Wildfeuer, Sherry. *Stella Natura*. Kimberton Hills, PA: Camphill Village Kimberton Hills. Reprinted by permission of the author.

85   Carpenter, Judith Perry. *Peacework Quilt: 365 Meditative Offerings*. Rockland, ME: Seafire Press, 2009. Reprinted by permission of the author.

89   Smith, Richard Thornton. *Cosmos, Earth and Nutrition*. (p. 137) East Sussex, UK: Sophia Books. 2009. Reprinted by permission of the publisher.

93   Wildfeuer, Sherry. *Stella Natura*. Kimberton Hills, PA: Camphill Village Kimberton Hills. 2010. Reprinted by permission of the author.

CHAPTER III

97   Tiwari, Maya. *Living Ahimsa Diet*. New York: Mother Om Media. 2011. Reprinted by permission of the publisher.

108   Haas, Elson M. *Staying Healthy With the Seasons*. Berkeley, CA: Celestial Arts. 1981. Reprinted by permission of the publisher.

CHAPTER IV

115   Johnson, Wendy. *Gardening at the Dragon's Gate*. New York: Bantam Books. 2008. Reprinted by permission of the publisher.

116 Adyashanti. *The Way of Liberation*. Campbell, CA: Open Gate Sangha, Inc., 2012. Reprinted with permission from the author.

124   Smith, Richard Thornton. *Cosmos, Earth and Nutrition*. East Sussex, UK: Sophia Books. 2009. Reprinted by permission of the publisher.

127   McIntyre, Anne. *The Complete Floral Healer*. New York: Sterling Publishing Co.. 2002. Reprinted by permission of the author.

128   Buchmann, Stephen L., and Nabhan, Gary Paul. *The Forgotten Pollinators*. Washington, DC: Island Press. 1996. Reprinted by permission of the publisher.

131   Richards, M.C. *Opening Our Moral Eye*. Hudson, NY: Lindisfarne Press. 1996. Reprinted by permission of the publisher.

CHAPTER V

139   Quote by Richo Cech, founder of Horizon Herbs. Reprinted by permission of the author.

143   Dawson, Adele G. *Herbs: Partners in Life*. Rochester, VT. Healing Arts Press. 2000. Reprinted by permission of the publisher.

McIntyre, Anne. *The Complete Floral Healer*. New York NY: Sterling Publishing Co. Inc.. 2002. Reprinted by permission of the author.

159 Wood, Matthew. *The Book Of Herbal Wisdom*. Berkeley, CA: North Atlantic Books. 1997. Reprinted by permission of the publisher.

174   Levy, Juliette de Bairacli. *Common Herbs for Natural Health*. New York: Schocken Books. First published by Schocken Books in 1974. Permission given by current publisher: Ash Tree Publishing.

176   Tiwari, Maya. *Living Ahimsa Diet*. New York: Mother Om Media. 2011. Reprinted by permission of the publisher.

195   Wood, Matthew. *Earthwise Herbal*. Berkeley, CA: North Atlantic Books. 2008. Reprinted by permission from the publisher.

215   Shiva, Vandana. *Staying Alive: Women, Ecology and Development*. Brooklyn, NY: South End Press. 2010.

Gretchen and Bob catching a swarm of honeybees in Avena's garden.

United Plant Savers

Planting the Future

Member of UpS
Botanical Sanctuary
Network

PHOTO CREDITS

A ll the photographs in this book have been taken in Avena Botanicals gardens except for a handful. I am especially grateful for the professional photographers G. Michael Brown, Mary Crowley, Susie Cushner, Lynn Karlin, and Stephen Orr for generously allowing their beautiful photos to be used in this book. Please visit their websites (listed below) to learn more about their work. A few of the photos were taken by Avena staff and friends (listed below). The rest were taken by me. The color images of the plants, pollinators, and people have brought the story of Avena's garden to life. Thank you to everyone for your contributions.

G. Michael Brown www.gmbrownphotos.com
Pages: 41 (male goldfinch), 46, 48 (warbler), 49, 51, 53, 54, 55, 60 (spider web), 78, 83, 86 (male goldfinch), 102 (frog on a lily pad), 106 (Golden-crowned kinglet), 210

Mary Crowley
Cover photo, 32, 77, 104

Susie Cushner www.susiecushner.com
Pages: frontispiece, 34, 59, 125, 130, 192, 249

Lynn Karlin
Pages: 86 (seeds and felco clippers), 117, 133, 183, 242

Stephen Orr, author and photographer of *Tomorrow's Garden*
Pages: 66, 87, 138, 248, 250

Many thanks to the following Avena staff and friends for your special photos.

Juliette de Bairacli Levy, dedication page, *Photo from the Juliette de Bairacli Levy Archive courtesy of her family*.
Bernard McLaughlin, p. 3: Photo from the McLaughlin Garden Archive.
Jennifer Neptune, p. 21: Alexandra Conover Bennett
Basket photos, p. 23: Jennifer Neptune

Buddha with chipmunk, p. 31: Pettina Harden

Yarrow, p. 68: Jean-David Derreumaux

Winter pond and willow tree, pp. 109 and 110: Jillian Kelsey-Rose

Deb in Sicily, p. 141: Tom Griffin

ADDITIONAL RESOURCES:

Allione, Tsultrim. *Feeding Your Demons: Ancient Wisdom for Resolving Inner Conflict*. New York, NY: Little Brown & C0. 2008.

Bowens, Natasha. *The Color of Food: Stories of Race, Resilience and Farming*. Gabriola Island, BC, Canada: New Society Publishers. 2015.

Buhner, Stephen Harrod. *Healing Lyme* (2nd edition). Silver City, NM: Raven Press. 2015.

Deming, Alison H. and Lauret E. Savoy (editors). *Colors of Nature: Culture, Identity and the Natural World*. Minneapolis, MN: Milkweed Editions. 2011

Gray, Beverly. *The Boreal Herbal: Wild Food and Medicine Plants of the North*. Whitehorse, Yukon, Canada: Aurora Borealis Press. 2011.

Hall, Matthew. *Plants as Persons: A Philosophical Botany*. Albany, NY: State University of New York Press. 2011

Kimmerer, Robin Wall. *Braiding Sweetgrass: Indigenous Wisdom, Scientific Knowledge, and the Teachings of Plants*. Minneapolis, MN: Milkweed Editions. 2013.

Kornfield, Jack. *Meditation for Beginners*. Boulder, CO: Sounds True. 2008, 2004.

Kumar, Satish. *Soil, Soul, Society: A New Trinity for our Time*. East Sussex, UK: Leaping Hare Press. 2013.

Phillips, Michael. *Mycorrhizal Planet: How Symbiotic Fungi Work with Roots to Support Plant Health and Build Soil Fertility*. White River Junction, VT: Chelsea Green Publishing. 2017.

Ray, Janice. *The Seed Underground: A Growing Revolution to Save Food*. White River Junction, VT: Chelsea Green Publishing. 2012.

Reynolds, Mary. *The Garden Awakening: Designs to Nurture our Land and Ourselves*. Cambridge, England: Green Books. 2016.

Shiva, Vandana. *Making Peace with the Earth*. Winnipeg, Manitoba, Canada: Fernwood Publishing. 2012.

Storl, Wolf. *Healing Lyme Disease Naturally*. Berkely, CA: North Atlantic Books. 2010

The Xerces Society. *Attracting Native Pollinators*. North Adams, MASS: Storey Publishing. 2011. www.xerces.org. Also check out Pollinator Partnership: www.pollinator.org

INDEX

ABOUT THE AUTHOR

DEB SOULE is an herbalist, biodynamic gardener, teacher, and author of *The Woman's Handbook of Healing Herbs*. Raised in a small town in western Maine, Deb began gardening at age 16. Her faith in the healing qualities of plants and her love of gardening led Deb to found Avena Botanicals Herbal Apothecary in 1985. Five years earlier, while in college, Deb lived in Nepal near three Tibetan monasteries and was deeply moved by the Tibetan people's commitment to ease physical ailments and mental and emotional challenges with plants, prayer, and other spiritual practices. Today, Deb tends three acres of medicinal plants using organic and biodynamic practices, teaches herb classes, studies pollinators, and consults with clients and health care providers. Her most recent inspiration is the Grow a Row Project. For more information: **www.debsoule.com**

ABOUT AVENA BOTANICALS

Founded in 1985, Avena Botanicals is known for the high-quality medicinal remedies and body-care products created directly on the farm by a small staff of people who love plants. We use certified biodynamic, organic, and wild-harvested herbs in our tinctures, non-alcohol glycerites, elixirs, infused oils, salves, crèmes, and herbal teas. Over 70% of the herbs used in Avena's products are gathered by hand from our certified biodynamic gardens and from nearby islands and meadows. What we cannot grow we purchase from other certified biodynamic and organic farms.

We are artisans and healers, working with our hands in similar ways that herbalists around the world have done for centuries. We are committed to creating the best traditionally crafted herbal remedies possible. This means knowing exactly when each herb we harvest is at its peak, then thoughtfully preparing our products by hand following time-tested recipes. For more information about our herbal remedies and biodynamic farm visit: **www.avenabotanicals.com**

ABOUT GROW A ROW

In my early twenties I volunteered for a domestic violence hotline. This experience greatly expanded my understanding of domestic violence, sexism, and misogyny. In my first book, *The Woman's Handbook of Healing Herbs*, I included herbal formulas for supporting a woman who is healing from an abusive experience or who has abuse memories resurfacing. Herbs and flowers work well alongside other healing modalities in supporting a woman's journey to wholeness.

Violence against women is prevalent worldwide. When I first learned about Eve Ensler's work in the Congo Region (www.vday.org) to assist women who have been raped and violated, I wept and said to myself, "How can I be of help?" The answer was to "grow a row of calendula."

Calendula is a remarkable herb for healing wounds and traumatized tissue and for filling the body and spirit with vitality and light. Used as an oil or salve, it heals all kinds of wounds, soothes and heals inflamed or infected vaginal tissue, and lessens scarring. As a tea, it offers much benefit as is described in the calendula section of this book (p. 181-184).

The Grow a Row Project

The Grow a Row project is a grassroots initiative (and not something organized by Avena Botanicals). Avena Botanicals is only hosting information about how people can get involved in The Grow a Row project on its website. My vision for this project is as follows:

Anyone who has a small or large organic garden can "grow a row" of calendula, collect the flowers regularly, and dry them. Store the dried flowers in a glass jar in a cupboard or in a clean brown paper bag.

Contact your local or county organization that serves women and children who have been raped or violated. Tell a staff person about the Grow a Row project and your wish to donate dried flowers for tea or healing baths or to donate calendula oil or salve that you make from your flowers. Print off and give information to the staff about the Grow a Row project and the educational sheet about calendula and its healing benefits. This exchange builds community among people and plants. If you are not located near an organization that is interested in giving calendula to their clients, then contact The Grow a Row Project through Avena Botanicals and we can give you the names of organizations that you can contact.

My hope is that this grassroots initiative will encourage women and men to grow and dry calendula, and will create more dialogue, community support, and understanding of the root causes of violence and oppression within families, schools, the health care system and government agencies. Planting seeds and sharing herbs is a way to bring about positive change.

Information sheets about growing, collecting, drying, and preparing calendula into teas, baths, oils, and salves are available at **www.debsoule.com**.

ABOUT THE HERBAL CLASSROOM

THE HERBAL CLASSROOM is Avena's non-profit, 501c(3), educational arm offering classes in herbal medicine, Ayurveda, meditation, and biodynamic and organic gardening. Indoor and outdoor classes occur in Avena's beautiful garden and octagon-shaped classroom with a mission to encourage students to use their hearts, hands, and minds in learning the healing ways of plants. The Herbal Classroom's programs train individuals and health care providers to work as educators, herbal practitioners, and gardeners in providing herbal knowledge and high-quality herbs to their community in responsible, earth-friendly, and community-minded ways. For information about our classes and monthly herb walks, visit: **www.herbalclassroom.org**